PRAISE FOR *THIRTY MILLION WORDS*

"Suskind's vision is empowering, her methods are surprisingly simple to execute, and the results have been proven to nurture children toward becoming stable, empathetic adults. Informative, exciting new data that confirms the significant benefits gained by talking to your child."
—*Kirkus Reviews*

"How empowering the realization that each of us can be our children's personal neuro-developer, that the sheer quality of our interactions can impact the attitude of inquiry and health of our kids. The practical application of Dr. Suskind's work is limitless; as a dad, and as (a 'gritty') someone invested in early child development, I'm excited to see how far it takes us."
—Steve Nash, president of the Steve Nash Foundation and two-time National Basketball Association MVP

"I thank Dana Suskind for opening doors to solutions and hope. The answer to the growing problem of social inequalities in our country is to make use of America's top two resources: its children and their parents. If we care about this country and the children who will live in it as adults, we have to make Dr. Suskind's advice become reality."
—Sandra Gutierrez, founder and national director of Abriendo Puertas/Opening Doors

"Grounded upon experience as a cochlear implant surgeon, informed by compelling social science research, and inspired by a profound commitment to children and families, this book helps all of us understand the importance of communicating abundantly, pleasantly, and responsively with infants and toddlers. Tell everyone you know to read it! Together, we can enrich language environments for generations to come, in all types of homes and communities. The book is a gem!"
—Ronald F. Ferguson, faculty director at the Achievement Gap Initiative at Harvard University

"Straight from the front lines Dr. Dana Suskind tells the story of the power of talk in helping children learn. Easy to read and with striking insights on every page, this book will not only make you think differently about how you parent but will give you the tools to help your child be at his or her best."
—Sian Beilock, professor at the University of Chicago and author of *Choke* and *How the Body Knows Its Mind*

"Suskind writes with clarity and authority to explain why parents need to talk to their infant children, and why some forms of communication are better than others. *Thirty Million Words* belongs on the shortlist of books that every parent, teacher, and education policy maker should read."

—Adam Alter, associate professor at New York University and author of *Drunk Tank Pink*

"Anyone who cares about children, anyone who cares about the future of this country, should read this book."

—Barbara Bowman, Irving B. Harris Professor at Erikson Institute

"A passionate, personal account of the power all parents have to raise thriving, successful children."

—Diana Mendley Rauner, president of the Ounce of Prevention Fund

"Prepare for a revolution. This book will make you cry, laugh, and deeply reflect on what we should be doing to give everyone a chance to succeed in life. As a scholar I was in awe, as a teacher I was dazzled, and as a father I was thankful to the authors of this book. When you pick it up, have a few hours to spare because you will not put it down. Unequivocal 5 stars."

—John A. List, Homer J. Livingston Professor at the University of Chicago and author of *The Why Axis*

"Dana Suskind hails others as heroes, but she is the true hero! She stepped out of the safe harbor of her role as a pediatric cochlear implant surgeon when she realized that helping deaf children hear was not enough to help them learn language. She takes us on her compelling and page-turning journey, navigating the best research in children's early learning, always sharing sound and extremely helpful examples of what we all must do to help children learn language and much more, always in joyful and loving ways."

—Ellen Galinsky, president of Families and Work Institute and author of *Mind in the Making*

"Dr. Suskind's work reveals that the greatest gift we can give our children is free. How empowering to understand that it doesn't take money to give our children advantages in the world, it takes words. Her research is vital." —Chris Nee, creator and executive producer of *Doc McStuffins*

"Three cheers for the promise of parents, prevention, and neuroplasticity!"

—Dr. T. Berry Brazelton, Clinical Professor of Pediatrics Emeritus at Harvard Medical School

THIRTY MILLION WORDS

BUILDING A CHILD'S BRAIN

TUNE IN,
TALK MORE,
TAKE TURNS

DANA SUSKIND, MD
BETH SUSKIND
LESLIE LEWINTER-SUSKIND

DUTTON
— est. 1852 —

DUTTON
— est. 1852 —

An imprint of Penguin Random House LLC
375 Hudson Street
New York, New York 10014

Copyright © 2015 by Dana Suskind and Beth Suskind
Penguin supports copyright. Copyright fuels creativity, encourages diverse voices, promotes free speech, and creates a vibrant culture. Thank you for buying an authorized edition of this book and for complying with copyright laws by not reproducing, scanning, or distributing any part of it in any form without permission. You are supporting writers and allowing Penguin to continue to publish books for every reader.

DUTTON—EST. 1852 and DUTTON are registered trademarks of Penguin Random House LLC

LIBRARY OF CONGRESS CATALOGING-IN-PUBLICATION DATA
has been applied for.

ISBN 978-0-525-95487-3

Printed in the United States of America
1   3   5   7   9   10   8   6   4   2

While the author has made every effort to provide accurate telephone numbers, Internet addresses, and other contact information at the time of publication, neither the publisher nor the author assumes any responsibility for errors or for changes that occur after publication. Further, the publisher does not have any control over and does not assume any responsibility for author or third-party websites or their content.

For Amelie, Asher, and Genevieve
DS

For Lily, Carter, and Michael
BS

For Bob and our wonderful family
LL-S

CONTENTS

THIRTY
MILLION
WORDS

CHAPTER I

CONNECTIONS

WHY A PEDIATRIC COCHLEAR IMPLANT SURGEON BECAME A SOCIAL SCIENTIST

Blindness separates me from things; deafness separates me from people.

—Helen Keller

Parent talk is probably the most valuable resource in our world. No matter the language, the culture, the nuances of vocabulary, or the socioeconomic status, language is the element that helps develop the brain to its optimum potential. In the same way, the lack of language is the enemy of brain development. Children who are born hearing, but in an austere language environment, are almost identical to children who are born deaf who have not received a rich sign environment. Without intervention, both can suffer the critical, lifelong effects of silence. On the other hand, children in a rich language environment, whether born hearing or given the gift of hearing via cochlear implants, can soar.

MY STORY

The irony is not lost on me that a pediatric cochlear implant surgeon is writing a book on the power of parent talk. Surgeons are known for many things; talk is not among them. Rather than words, we are defined by our hands, our dexterity in the operating room, and our ability to identify problems and find solutions. To a surgeon, there is nothing more gratifying than when the puzzle pieces fit just so.

Cochlear implantation, allowing a child born deaf to hear, is an extraordinary example of all those components. Coiled two and a half times around the snail-shaped cochlea, the organ where the nerve part of hearing begins, a cochlear implant successfully skips over the defective cells, the point along the hearing pathway where sound had come to a screeching halt, going directly to the *acoustic,* or *hearing,* nerve, the superhighway that connects the ear to the brain. The amazing result is that a child born into silence now has the ability to hear, to talk, and to meld into the world both educationally and socially. The cochlear implant is a puzzle piece that fits, just so, a miraculous solution to total deafness.

At least, that's what I thought.

In medical school, it was the brain, not the ear, that captured my imagination. The brain seemed a profound mystery holding the key to all the unanswered questions about life. My dream was to be a neurosurgeon, fixing with my hands some of the most important and vexing issues facing humanity.

My first neurosurgical case in medical school did not, however, go smoothly. The chief of neurosurgery, Dr. R., had invited me to "scrub in" on a meningioma resection, the removal of a benign brain tumor. We'd been writing a textbook chapter on meningioma resection and he thought it might be helpful if I actually saw one. When I came into the operating room, Dr. R. gestured me toward the operating room table, where a shaved head, stained with the yellow and red of Betadine and blood, confronted me. Inside a large gap of missing skull, a grayish gelatinous mass pulsated rhythmically, as if trying to escape its bony confines. The patient's torso had disappeared completely, like a magician's assistant's, under long blue drapes.

As I walked toward the patient, I suddenly became aware of my own pulsations. Could this blob of overly congealed gelatin really be the epicenter of who we are? Dazzling lights crisscrossed my vision and I could barely register what Dr. R. was saying. The next thing I knew, I was being lowered onto a chair by one of the surgical nurses. Humiliating? You bet!

But that was not the reason I did not go into brain surgery. Ultimately, that was more a fantasy-meets-reality decision.

"When the air hits the brain, you're never quite the same" was a common saying in neurosurgery back in the 1980s. At that time, surgery on the brain often left patients severely debilitated, though alive. In the intervening years, of course, things have changed for the better, but my own experience prompted me to think of other ways of working with the brain. And, in a circuitous way, I did: the ear. Under the guidance of my extraordinary mentor, Dr. Rod Lusk, during my fellowship at Washington University in St. Louis, I learned the skills necessary to help ensure the success of cochlear implantation.

Cochlear implantation is, to me, one of the most elegant of surgeries. Performed under a high-powered microscope that magnifies the inner ear from the size of a tiny pea to the size of a quarter, it relies on small, precise instruments that match its small, precise movements. I operate with the room lights off, the single light beam from the microscope a spotlight on the star of the show, the ear. The microscope's penetrating beam has been said to cast an almost romantic halo around the patient and surgeon. And while many surgeons operate with music, I prefer my operating room to be quiet and calm, with only the hum of my drill as backdrop to my concentration on the surgical procedure.

My decision to become a pediatric head and neck surgeon specializing in cochlear implantation was serendipitous. Because the confluence of two historic medical events was about to usher in a golden age for children born deaf.

In 1993, the National Institutes of Health recommended that all newborns undergo a hearing evaluation, the universal newborn screening, before leaving the hospital. This astute public health initiative plummeted the age of the diagnosis of deafness from three years to three *months*. No longer could parents and pediatricians rest comfortably with "He's just a slow talker" or "Her older brother does all her talking," when, in fact, the child was deaf. But its significance was increased exponentially because it coincided with the development of a neurologic miracle, the cochlear implant. The possibility of changing the life course for millions of deaf children had arrived.

THE COCHLEAR IMPLANT

The brain and neural structures in the human body are generally unforgiving. From cerebral palsy to strokes, from spinal cord injuries to football-related head trauma, "making better" rather than "correcting" tends to be the medical dictum. Hearing loss is a spectacular instance where something can, in fact, be done.

In 1984, the first single-channel cochlear implant for adults, allowing sound detection and some awareness of voice, although not "hearing" as we know it, was approved by the FDA. This was followed, in 1990, at about the same time as the new recommendations for universal newborn screening, by a new multi-channel cochlear implant with complex speech processing ability, approved for young children. For the first time in history, a child born deaf would be able to hear at an age *when the brain pathways for language were being created.*

It's important to understand why the timing of these two coinciding events was so critical. By the end of age three, the human brain, including its one hundred billion neurons, has completed about 85 percent of its physical growth, a significant part of the foundation for all thinking and learning. The development of that brain, science shows us, is absolutely related to the language environment of the young child. This does not mean that the brain stops developing after three years, but it does emphasize those years as critical. In fact, the diagnosis of hearing loss in babies had often been called a "neurologic emergency," essentially because of the expected negative impact on a newborn's development.

The importance of early screening coinciding with cochlear implantation for children cannot be overstated. If they had *not* taken place simultaneously, if, for example, diagnoses of deafness came at a later age, and the cochlear implant had been placed in older children's ears, the cochlear implant may have been deemed a fabulous piece of technology but not much more, certainly not the game changer that it is. That's because successful cochlear implantation requires *neuroplasticity,* the ability for a brain to develop with new stimuli. And although neuroplasticity for learning language may occur, to some degree, at all ages, it is integral to the young brain from birth through about three to four years of age. Exceptions are those who have gone deaf *after* having learned to talk and whose brain's language pathways are already established. Those who are born deaf and receive implantation at a much later age will hear sounds, but rarely will they gain the ability to understand their meanings.

I soon learned, however, that even when cochlear implants are put in place at an optimum time, there are other factors that may preclude their success.

THE ADVANTAGE OF STARTING SLOWLY

The University of Chicago is an island in a sea of inequality on Chicago's South Side. Adding to the overwhelming social and economic challenges facing many families on Chicago's South Side, before I began my cochlear implant program there was the added barrier of communication between children born deaf and their families. This presented both a remarkable opportunity and an extraordinary challenge for me and my incredible, dedicated cochlear implant team. It would also prove to be the experience that would entirely change the direction of my thinking and my career.

When I was an infant in the late 1960s, at the height of the civil rights conflict, my mother, a social worker, took me to work with her in inner-city Baltimore. I slept in a room near her office with someone sitting outside of the door to tell her when I woke up. Later that year, sent to Peru to do a study of the possibilities for creating infant care centers in the *barriadas* circling Lima, she would sometimes carry me through the hills on her back in an aluminum-sided baby carrier, an "in," she said, with skeptical inhabitants who had never seen a foreigner do that. Much later she told me that whatever she did, anywhere, never came close to how much she learned, especially about the wealth of untapped potential in people who never had a chance. It's the same experience I've had with my own patient population. Little did I know, when I began this journey, that one of the biggest impacts of my work would be, in fact, on me.

My cochlear implant program at the University of Chicago started slowly. Patients weren't, as I had thought they would be, lining up at my door like at a big sale day at a shopping mall. But it was the slow start that gave me a crucial perspective I might otherwise have overlooked.

Because there were so few, I tended to each patient as if he or she were my own child, noticing each milestone, a first smile, a first step, with all the pride of a parent. I was present at every activation, the moment a child's cochlear implant is turned on and sound is heard for the first time. And, like a parent, while I was rejoicing successes, I was also agonizing when things were not as they should have been.

The problems I saw bothered me tremendously; lagging responses to first hearing sound, lack of reaction to hearing their names, slowness in saying a first word or reading a first book. Adding to this burden was the fact that the profound differences I saw occurred in children who had looked very similar to the others at the outset. The path to find out why would eventually lead me into the world of children born hearing.

The truth is, at one time I would have likely dismissed my observations of the children as non-science, interesting anecdotes at best. For me, as for many others in academia, science became "true" science only when the numbers were big enough to prove or disprove something, a sample size with "power," as we say. But I soon came to realize that the power of numbers, disregarding the significance of individual experience, can obscure important insights.

ZACH AND MICHELLE

Zach was my second cochlear implant patient; Michelle, my fourth. Both diagnosed as profoundly deaf at birth, they were strikingly similar in many ways. Both showed similar innate potentials, both had mothers who loved them and wanted them to live in the speaking world, and both were recipients of some of the most powerful technology science had to offer. But that's where the similarities ended. Same potential, same surgery, but very different outcomes.

I could never have learned what I learned from Zach and Michelle in any medical textbook. It is not just that my experience with them made me become aware of the limits of technology; it also made me acknowledge a force whose potential impact I may have always known but had failed to recognize, a force that irrevocably impacts the arcs of all of our lives.

Zach

Zach was about eight months old when his parents brought him to see our team, a peanut with hair so light you barely realized he had any. He smiled easily; his blue eyes, the color of a clear sky, watched our every move. His deafness had been a shock to his parents. No one in his family had hearing loss except one cousin who had gotten hearing aids in his sixties. His sister, Emma, two years older than Zach, had normal hearing and was the definition of the chatty older sister. But although his parents had had no contact with anyone who was deaf, they knew what they wanted before they entered my office.

Zach's parents had educated themselves. No-nonsense, quietly determined, they were aware that there were communication choices and they effectively let us know their goal: to have Zach be part of the hearing and speaking world. Zach had already been wearing hearing aids almost since his diagnosis and, astonishingly, while parents often battle children to keep them on, Zach wore his easily, his tiny ears flipped over like palm trees in a hurricane by their weight.

Zach's parents were proactive in other ways, as well. From the beginning they had a therapist come to their home to work with them and Zach on techniques to enhance his language development. They even started to learn sign language because they wanted to make sure Zach would be able to communicate no matter the mode. As a result, sign language was already the connection between Zach and his family.

From the beginning, Zach's parents knew that cochlear implantation was a possibility. The problem for Zach was the timing. His auditory brainstem response (ABR) test, done when he was an infant to determine hearing, had come back "no response," a flat line streaming across his ABR tracing with no pretty neural peaks indicating a brain's response to sound. The requisite hearing aid trial had also failed; Zach had the most profound form of deafness that exists. How were hearing aids going to make a dent when ninety decibels, the sound of a motorcycle racing by, didn't register a blip in Zach's brain even *with* hearing aids? Nonetheless, Zach's parents, who never gave up, had Zach fitted with hearing aids in the hope that he was the rare exception and they would actually work. What else were they to do for a year while they waited to comply with FDA guidelines that approved implantation only for children twelve months or older?

Always proactive, Zach's mother, recognizing from the outset that the hearing aids weren't working, sought the answer on her own. When he was a baby, she would lay Zach on her chest and place his tiny hands on her voice box, hoping he would connect the vibrations of her sweet lullabies to sound. In the same spirit of finding a solution, when she brought Zach to see me there was no question of their intent to have cochlear implantation. His first birthday, his parents decided, would be his "hearing" birthday.

While implantation is the first step, the true "hearing birthday" is actually at the moment the cochlear device is activated. A very dramatic moment, it is invariably followed by, "Honey, Honey, do you hear Mommy? Mommy loves you so much," then, when it's successful, the startled expression of the child followed by a smile, laughter, or even crying. It is an extraordinarily moving experience. See for yourself. Simply search YouTube for "cochlear implant activations" and wait for the tears.

On Zach's real "hearing birthday," both he and his parents were cool and relaxed. So relaxed, in fact, that they didn't even video record the occasion, one of the few regrets his mother has.

Like all first birthdays, of course, cochlear implant activation day is only the beginning step toward the goal of speech. And while parents often believe, even though they've been counseled to the contrary, that the ride from activation to spoken language is smooth sailing, a few days at the most, it isn't. Just like hearing newborns, newly implanted children must spend about a year soaking in, and learning to understand, the sounds in their world. Not always that easy. Zach, before implantation, could not hear a motorcycle roaring by; after implantation, he could hear the quietest whisper. Nonetheless, while he heard the sounds, his brain didn't have a clue what they represented. Which is what he, and all implanted children, have to learn before they can begin to speak.

Zach's world at home was filled with talking, reading, and singing. But while his parents swore that he was progressing beautifully, this was never apparent to me. During clinic appointments, even bribery with toys, stickers, anything that might prompt a word, failed. So it was only by humorous accident, when he was three years old, that I discovered that, yes, Zach really could talk.

A violin recital, *The Gift of Sound,* was being performed by members of the Chicago Symphony Orchestra in honor of our implant program, with many of our program's families in attendance. While music swept through our hospital lobby, people milled around, helping themselves from a long table piled high with cookies and other treats. And it was from this table that I received absolute confirmation that Zach could speak. Because, suddenly, somewhere between the brownies and the cookies, in the middle of Paganini or perhaps Beethoven, came high-pitched child's laughter and a loud, gleeful exclamation: "Ewww! Daddy faaaarted." And it was then that I knew all was going to be fine for Zach.

Zach is now a mainstreamed third grader in public school. The only outside educational services he receives are from a hearing specialist who ensures that his cochlear implant device is in good working order. He learns at grade level, including reading and math, plays with his friends, fights with his older sister, and gets no special treatment from his no-nonsense, loving parents. He is just a nine-year-old boy with intelligence, spirit, and every indication that he will fulfill his potential. His future is not defined by his hearing loss. He is lucky in many ways.

If Zach had been born twenty years earlier, in 1985 rather than 2005, his hearing loss *would* have defined his future. While there are many ways to live a happy and fulfilled life, even without hearing, the advent of cochlear implantation transformed Zach's education and career choices. This is largely because the ability to hear has an impact on the ability to read and, in consequence, to learn. The domino effect, over a lifetime, is evident. In studies done on adults born deaf and educated solely with sign language, the average literacy level in the past was fourth grade; one-third of deaf adults are functionally illiterate.

These statistics are not, of course, representative of those fortunate to live in homes rich with the language of native or skilled signers. Neither do they discount those in the deaf community who have achieved beautifully in the arts, in the sciences, in life. When there *is* lack of achievement, however, it is often related to the fact that about 90 percent of deaf children are born to parents who, while loving, cannot communicate with sign language, so that during the child's critical early years, when optimum neuroplasticity permits brain development, the necessary language environment is inadequate.

Compare this to Zach. Born deaf, yet reading at grade level in the third grade, which is often considered the predictor of long-term academic success, Zach is evidence of the perfect alignment of the stars of parental initiative, technology, and medical policy.

Michelle

A rich language environment "is like oxygen. It's easy to take for granted until you see someone who isn't getting enough."
—With apologies to Nim Tottenham for liberties with her wonderful quote

Seeing the puzzle pieces fit together perfectly allows one to see the beauty of possibility. It also puts into stark relief what happens when a puzzle piece is missing. It's here that Michelle's story and my turning point begin.

Michelle at seven months looked like a Japanese anime heroine; her crystal-blue-eyed gaze was soulful, intelligent, and entrancing; her laughter, joyful. Like Zach, Michelle had been born without hearing but with all the potential in the world. The puzzle piece she was missing was subtle and at first I didn't really know it existed. In fact, if Michelle had come before Zach, I would likely have either accepted her lag in progress as technology's limitations or simply attributed it to the fact that some "just don't benefit." But Zach had already set the bar, and what was happening to Michelle after her cochlear implant did not come close to my expectations of what *should* be happening.

Michelle's father had a moderate hearing loss that was correctable with hearing aids and attributable to Waardenburg syndrome, a genetic condition that affects, among other things, hearing. Like Michelle, who also had Waardenburg syndrome, he had widely spaced blue eyes and normal intelligence. Our team counseled Michelle's mother, Laura, at length. It was clear that as much as she loved her daughter, the weight of her world, including being unemployed with little money, and that now included a child with disabilities, was a heavy burden. It was decided that we would first attempt hearing aids, although I felt, with Michelle's hearing loss, they would probably not be enough. If they were not successful, we all agreed that cochlear implantation would be the next option. Soon after Michelle received the hearing aids, however, Laura moved away, and our professional role with Michelle ceased. When they returned a year later, Laura confirmed that the hearing aids were not working and that she had decided to follow our original suggestion to have cochlear implantation.

I clearly remember Michelle's "hearing birthday" at about two years of age. At that time, we celebrated activation by giving the patient a cupcake and a brightly colored balloon. It was, after all, a festive occasion; although, in Michelle's case, a moderate one. When the cochlear implant was activated, Michelle simply continued eating her cupcake, showing very little response. But "very little response," is quite different from "no response." Both Michelle's mother and I were delighted; Michelle, it seemed, could hear, which meant she could learn to talk.

Michelle's hearing, with the implants, was eventually evaluated as in the normal range. The audiologist and speech therapist both referred to her as a "sponge," easily responding to whatever they were trying to elicit from her. But something else was also apparent. While she responded to *sound* in the testing booth, she neither used nor seemed to understand speech. Her mother had noticed this at home, as well. Ultimately it was acknowledged that while Michelle could hear sounds, she did not understand their meanings, nor did she seem to be able to *learn* to understand their meanings.

This was very concerning to all of us who worked with Michelle professionally, including her therapists and the audiologist. During our implant team meetings, ways to support Michelle and her mother were discussed, including efforts to accelerate Michelle's language development by exposing her to more sign and spoken language. But none of these interventions was successful. Unlike Zach, who simply became silent in front of me, Michelle was truly silent, her problem far more serious and complex.

What had gone wrong? I had provided the gift of hearing to two deaf children. Why hadn't that been the complete answer to speaking and learning and integrating into the world? What were the salient differences that separated the outcomes of Zach and Michelle? The answer took me out of the world of the deaf into the wide world of all of us. Because the factors that differentiate Zach's and Michelle's abilities to learn are essentially the same that determine reaching learning potentials for all of us.

THE SIGNIFICANT DIFFERENCE

The reading level in third grade generally predicts the ultimate learning trajectory for all children. In the third grade, Zach is learning and functioning at grade level.

Michelle is also in the third grade, although in a Total Communication classroom. Even with a working cochlear implant she functions with minimal spoken language and only a basic grasp of sign language; the hope for true spoken language a distant dream. In addition, her third-grade reading is barely at the level of a kindergartener, a predictor of her life to come.

Why had the miraculous promise of a cochlear implant passed by this bright little girl with so much potential?

It turns out that what had gone wrong goes wrong more often than I had realized. This fact became starkly apparent when my team and I toured Chicago schools' hearing loss classrooms so that we could better understand the educational landscape our patients entered. The classrooms we visited were divided into "Oral," where spoken language was the primary form of communication, and "Total Communication," where, despite its name, sign language was the primary form of communication, with some spoken language. I, of course, had been sure that all of the children I had implanted early would be in the exclusively Oral classrooms. I was very wrong.

The Total Communication classroom had nine students in a semicircle of desks facing the teacher, who was signing to them. The silence was overwhelming.

And then I saw Michelle, whose blue eyes absolutely identified her. I went over and gave her a hug. Michelle, having no idea who I was, looked up at me with a confused, shy smile. No longer the vibrant little toddler I had first known, her sparkle seemed to have faded completely. With reason. Her teacher shared with me the hardships Michelle had gone through, including coming to school with no lunch, wearing dirty clothes and, most important, an inability to communicate well in either spoken or sign language. When I looked at her lovely face, it was hard to say whether I was seeing the tragedy of deafness or the tragedy of poverty. Without question, however, I knew that I was seeing the tragedy of wasted potential.

Two babies had come to me with very similar potentials but with very different outcomes. Yes, their backgrounds were entirely different, but socioeconomic status had never stopped a child from learning to talk. As a surgeon who had put so much faith in this magical puzzle piece that "fit just so," who had extoled a golden age for children born deaf, I was devastated, humbled, and, above all, newly determined.

Taking the Hippocratic oath meant that my obligation didn't end when I finished operating; it ended when my patient was well. I knew, absolutely, that it was time for me to step outside the comfortable world of the operating room.

THE UNIVERSITY OF CHICAGO
A WONDERFUL HOME

At the University of Chicago, I am surrounded by incredible medical and social scientists, including Nobel laureates, many in search of solutions to our world's most vexing problems. It's important to acknowledge that I had never been one of them. My world was the operating room. My ultimate goal was the implanting of cochlear devices to bring hearing to deaf children, making sure they were working properly, giving a hug and a kiss, and assuming all would be well.

So much for assumptions.

The life we are born into is simply the luck of the draw. No infant emerging into this world knows what's in store; there are no checklists for what you can expect in life, no menu that says one from column A, two from column B. And yet, from day one those factors, over which we have no control, have an indelible effect on our entire lives. In addition, while there is no socioeconomic relationship to being loved as a child, or in having parents who want you to be happy and fulfilled, or to having enormous potentials, there are definitely socioeconomic factors that relate to educational attainment, health status, and disease outcome.

This I learned from stepping out of the operating room and into the wide world of social sciences.

The terms "health disparities" and "social determinants of health" relate to the fact that in virtually every disease, from cancer and diabetes to obscure problems such as presbyosmia, the age-related loss of smell, significantly worse outcomes occur in those born into poverty. What I began to understand from my wonderful and esteemed colleagues at the University of Chicago was that Michelle's problems were related to the world into which she had been born. But knowing this provoked other questions. Were we saying that there was no solution? Do we say that's that and go on to another, more promising patient? Anyone who has read Emma Lazarus's poem on the Statue of Liberty, "Give me your tired, your poor . . . the wretched refuse of your teeming shore," knows that the next step is not accepting the historical "inevitable." The next step is changing the "inevitable" by finding a solution.

For a surgeon, trying to find a solution to a social problem means having to leave the familiar confines of the hospital and the operating room, a bit like planning a trip to the moon. On my way to work, I had often crossed the beautifully landscaped historic stretch of Gothic carved stone architecture known as the "Quad," where University of Chicago research scientists, otherwise known as the "Giants," do their thinking, teaching, and studying. And it was in that community of social scientists committed to finding out the intricacies of human behavior that I would begin to understand why Michelle's language had never developed as it should have and, most important, how I might have helped.

Professor Susan Levine and Professor Susan Goldin-Meadow, otherwise known as "the Susans," are University of Chicago professors of psychology, colleagues, longtime friends, and next-door neighbors. For four decades they have been working together to understand how children learn language. They opened my eyes, or rather, provided me with a new lens with which to see the world, especially the world of language acquisition.

I audited Susan Goldin-Meadow's undergraduate class Introduction to Child Language Development during a bitter Chicago winter. Often running late from my clinic, I would rush through the Quad, a heavy down coat covering my white lab coat, which, in turn, was covering my green scrubs. In the antiquated auditorium a steep incline of desk chairs funneled down to the lecture podium. As if proximity would compensate for the neurons no longer firing as rapidly as those of the college students surrounding me, I usually sat in the front row, listening as the students enthusiastically debated the Chomsky-versus-Skinner opposing theories of language acquisition. Was Chomsky correct that each of us is born with a "language acquisition device," an internal hard drive with the grammatical rules of language already preloaded into our brains? Was learning language our innate biological destiny? Or was Skinner correct when he hypothesized that learning language was not innate, but simply a phenomenon of adult reinforcement, eventually guiding children to acceptable language patterns? These were questions far from the cut-and-sew setting of the operating room, but they were now absolutely part of my world. I was acutely alert, waiting for the insights I needed to help the children I cared for.

HART AND RISLEY

I don't think I'd ever heard of Hart and Risley before Susan Goldin-Meadow's class and when I first heard their names in her class, I'm sure I had no idea of their ultimate importance to me. Child psychologists at the University of Kansas in the 1960s, Betty Hart and Todd Risley wanted to find a way to improve the poor academic achievement in low-income children. The program they designed, which included intense vocabulary enrichment, initially seemed to work. But when the children were tested before entering kindergarten, the positive effects had faded. Hart and Risley's determination to find out why resulted in a landmark study that was pivotal in our understanding the importance of the early language environment in a child's long-term learning trajectory.

But what makes Betty Hart and Todd Risley extraordinary is not simply the results of their study, but the fact that they did the study at all. The conventional wisdom of the time was that if you do well it's because you're smart; if you don't do well, it's because you're not. End of discussion. The differing trajectories of children born into poverty versus those born into more affluent families had long been accepted as immutable fact. Rarely were causes sought, because everyone knew why: genetics.

Hart and Risley changed that. In their groundbreaking study, they found another answer to the pivotal question "why?" The language environments of young children born into poverty, their study showed, were very different from the language environments of children born to more affluent families, and those differences could be correlated to later academic performance. In addition, while their study showed that children in lower socioeconomic homes heard far less language than their higher socioeconomic counterparts, quantity was not the only difference. Hart and Risley also found significant differences in quality, that is, *what types* of words were spoken and *how* they were spoken to a child. Finally, confirming that language exposure, not socioeconomic status, was the salient difference, Hart and Risley found that no matter how well or how poorly the children did academically, the early language environment was the significant factor. It all came down to words.

Because of Hart and Risley, the importance of the early language environment began to be understood: that the words a child heard, both the quantity and the quality, from birth through three years of age could be linked to the predictable stark disparities in ultimate educational achievement.

WHERE IT ALL COMES TOGETHER

The children in the Hart and Risley study were born hearing, yet they were no different from children born deaf who acquire cochlear implants. Children who receive cochlear implants and are in homes where the language is rich do well; those who receive cochlear implants but are in homes without adequate language do not do as well. I began to understand, thanks to the work of many dedicated scientists, that it takes more than the ability to hear sounds for language to develop; it is learning that the sounds have meaning that is critical. And for that, a young child must live in a world rich with words and words and words.

I had given all of my patients the same ability to hear, but for those children born into homes where there was less talk, less eliciting of response, less variation in vocabulary, the meaningful sounds needed to make those critical brain connections were not sufficient. The cochlear implant, as incredible as it is, is not the missing puzzle piece. Rather, it is simply a conduit, a pathway for the essential puzzle piece, the miraculous power of parent talk, a power that is the same, whether a child is born hearing or has acquired hearing via a cochlear implant. Without that language environment, the ability to hear is a wasted gift. Without that language environment, a child will be unlikely to achieve optimally.

I believe that every baby, every child, from every home, from every socioeconomic status, deserves the chance to fulfill his or her highest potentials. We just have to make it happen.

And we can.

And that is what this book is about.

CHAPTER 2

THE FIRST WORD

THE PIONEERS OF PARENT TALK

*Never doubt that a small group of thoughtful,
committed, citizens can change the world. Indeed,
it is the only thing that ever has.*

—attributed to Margaret Mead

In 1982, two perceptive social scientists from Kansas City,
Kansas, Betty Hart and Todd Risley, asked a very simple
question: Why had their innovative program to help prepare
high-risk preschoolers for school failed? Designed to enhance
the academic potential of children by intensively increasing their
vocabulary, it seemed a perfect solution to a prevailing problem.
But it was not.

The initial results of Hart and Risley's project had been encouraging. Aware of the importance of language in a child's
scholastic success, Hart and Risley had included a rigorous vocabulary component in the intervention. Its goal was to boost
the children's lagging vocabularies so they would enter kinder-

garten on a par with better-prepared peers. Initially, Hart and Risley did observe a promising "spurt of new vocabulary words . . . and an abrupt acceleration in . . . cumulative vocabulary growth curves." But while the children gained vocabulary as a result of the intervention, it was soon apparent that their actual learning trajectories remained the same and, by the time they entered kindergarten, the positive effects had disappeared and they were no different from the children who had not attended the intensive preschool program.

Hart and Risley's hope, like many of their generation's, was to break the "cycle of poverty" through preschool education. Active participants in Lyndon Johnson's War on Poverty, they were exemplary examples of their time, aiming "not only to relieve the symptom of poverty, but to cure it and, above all, to prevent it."

Their initial search for answers began in 1965. While much of the United States was erupting in race riots and civil unrest, Hart and Risley met with University of Kansas colleagues to design a project aimed at drastically improving the academic achievement of impoverished children. Called the Juniper Gardens Children's Project, it began in the project's headquarters in the basement of C.L. Davis's liquor store. Its ultimate design combined "community action with scientific knowledge" and included a rigorous, vocabulary-intense curriculum aimed at increasing the children's school readiness and academic potential.

Grainy evidence of this project still exists in a 1960s YouTube video: "Spearhead—Juniper Gardens Children's Project." In it, a young Todd Risley, in skinny black suit and skinny tie, walks purposefully into their "laboratory" preschool. In one of the classrooms, a young, smiling Betty Hart sits schoolmarm style on the floor, reading responsively with a circle of four-year-olds. The soaring tone of the film parallels their hope that "pressing social problems could be solved by improving everyday experiences." The video ends with a crescendo and a dramatic voice-over: "This is one small inroad, a spearhead at Juniper Gardens, where community research seeks to overcome the obstacles which separate the children of our deprived communities from the abundance of the rest of the nation."

The failure of the Juniper Gardens Children's Project could easily have been attributed to the prevailing answers of the times: genetics or some other unalterable factor. But Hart and Risley were not blithe acceptors of "conventional wisdom." Refusing to accept their results as the final answer to a widespread problem, they insisted on finding out *why* it failed. The study they designed opened the door to understanding that the prevailing thought on why children fail was flawed and that there was, in fact, the potential for changing what had been perceived as unchangeable.

ROMANTICS

Steve Warren described Betty Hart and Todd Risley as "romantics." Currently a professor at the University of Kansas, Steve Warren was a young graduate student in the 1970s when he first met Betty Hart and Todd Risley.

"Romantics," he said, but not "head in the clouds" romantics. Undeterred by the prevalent philosophy of ascribing the failures of the War on Poverty to genetics, refusing to abandon populations written off by society, they became detectives, asking questions that might lead to solutions to perpetual problems.

The two questions they asked were:

> What happens in the life of a baby and a young child during the 110 hours of being awake each week?
> How essential is what happens during that time to the eventual outcome of the child?

Which led to an incredible realization.

"There was [absolutely] no literature [on the daily lives of babies] . . . none . . . which was shocking when you think about it."

While there may have been some literature, there seemed to be little incentive, until Betty Hart and Todd Risley began their work, to pursue answers or solutions.

REVOLUTIONARIES, AS WELL

Hart and Risley's insight into the role of early language exposure in a child's ultimate academic achievement was an incredible step forward in social thought. The famous "war of words" between Noam Chomsky and B.F. Skinner, debating the question of language acquisition during the same period, didn't even hint at language exposure as a factor.

An intellectualized debate, the "war of words" juxtaposed Chomsky's theory of genetic prewiring of the human brain, or "nature," with B.F. Skinner's "operant conditioning," that is, negative versus positive reinforcement, or "nurture," as the predominant factors for acquiring language. Most incredibly, while Skinner was the "nurture" part of the argument, exposure to language via parental input was not even mentioned in his theory. Instead, Skinner's "operant conditioning" rested on the theory that a child's language acquisition resulted from reinforcements similar to Pavlov's "rat with a lever" reward-versus-punishment routine.

Chomsky theorized, on the other hand, that humans had a "language acquisition device" genetically prewired within their brains. He believed that brain "encoding" explained the early, rapid acquisition of language in young children. Dismissing B.F. Skinner's hypothesis as "absurd," he questioned how the complexity of grammar that children acquire in such a short period of time could be explained by a simplistic theory of reward and punishment.

The general acceptance of Chomsky's theory mirrored the broader acceptance of the importance of heredity in what happens to humans. As a result, interest in and support for exploring disparities in language outcomes were rare. Research in language acquisition was done primarily with infants and children from middle-class families, with findings then generalized to all children. There was little attempt to examine variations in development. While the debate continues today, as evidenced by the heated discussions I witnessed while auditing Professor Susan Goldin-Meadow's undergraduate course on child language development, it is Hart and Risley who deserve credit for initiating awareness of the importance of early language exposure in the development of the intellect.

Todd Risley
Do Good And Take Data

While both Hart and Risley believed that science's role was "for the social good that [it] could produce" and to help "develop answers to serious human problems," in many ways Betty Hart and Todd Risley were polar opposites. It may have been their dissimilarities, in fact, that led to their successfully taking a very innovative, not totally accepted idea, and turning it into a world-famous landmark study.

"Applied behavioral analysis" refers to applying what science tells us about human behavior toward solving social problems. Todd Risley, a developmental psychologist, was one of its founding fathers, dedicating his professional life to understanding how human behavior could be shaped through interventions.

"Todd's genius," said his lifelong colleague James Sherman, "was in seeing through the . . . tangle of vines . . . to the essence of the problem" in order to solve it. In other words, Risley paved clear paths through the behavioral labyrinth.

Betty Hart
The Perfect Partner

Betty Hart, said Steve Warren, was "a unique genius." Reserved, shy, wearing large glasses that overpowered her thin face, she had been Todd Risley's graduate student in the 1960s. Her escalation to colleague did not change their relationship; even after becoming his research partner, Betty Hart called Todd Risley

"Dr. Risley." But her mild academic appearance belied a dogged perseverance for detail and accurate data, traits that propelled their study from a vision into a reality. In 1982, Todd Risley left Kansas City, returning to "Risley Mountain," his family's homestead in Alaska for four generations, to become professor of psychology at the University of Alaska in Anchorage. When he left, the day-to-day burden of the study fell on Betty Hart.

THE STUDY

Forty-two families from all socioeconomic strata were selected to participate in their study. Their children were followed from about nine months of age to three years. Socioeconomic levels were determined by family occupation, maternal years of education, highest educational level of both parents, and reported family income. This resulted in thirteen "high" SES families, ten "middle" SES families, thirteen "low" SES families, and six welfare families participating in the study. One prerequisite for all families was stability, or "permanence": Did the family have a telephone? Own a home? Did they expect to stay in one place for the foreseeable future?

Originally fifty families had been selected for the study, but this number was reduced when four families moved away and four "missed enough observations that their data could not be included in the group analysis." In hindsight, these families may have actually represented an important subset for their data analysis.

Recognizing that they were starting from scientific scratch, Hart and Risley decided to record absolutely everything.

"Because we didn't know exactly which aspects of [a child's total daily experience] were contributing to . . . vocabulary growth, the more information we [gathered] . . . the more we could potentially learn."

The study took three years. Once each month, during those three years, for one hour each session, a study observer audiotaped and took notes, recording everything "done *by* the children, *to* them, and *around* them." The team Hart and Risley assembled was so committed to the research that, according to records, no one took a single day of vacation during the entire study. After three years of painstaking, detailed observation and an *additional* three years of data analysis, Hart and Risley were "finally ready to begin formulating what it all meant."

In our age of split-second, instant-answer computers, the fact that Hart and Risley's team had to spend three additional years, an additional twenty thousand work hours, analyzing the data, seems almost unbelievable.

Most of the work fell on Betty Hart. Todd Risley once referred to Betty Hart as a "foreman," but, to me, Betty Hart is an unsung hero. Her assiduous devotion to accuracy, both in data collection and analysis was pivotal to the successful completion of one of the most important studies in early childhood development. While Hart and Risley are proof that there is rarely true geni*us* without "*us*," I also believe that without Betty Hart there would have been no completed study at all.

Although Hart and Risley's research was aimed at finding differences, their most extraordinary finding was the similarities between families from the different socioeconomic strata. "Development," said Hart and Risley, "made all the children look alike." When "we saw a child in one home starting to say words, we knew we would see all the other children [doing the same thing]."

Parents were also similar. "Raising children made all the families look alike" as parents "socialized their children to a common cultural standard." "Say thank you." "Do you have to go potty?" All the parents, from every socioeconomic group, according to Hart and Risley's report, wanted to do the right thing, each struggling at the hard job of raising autonomous beings.

"Our surprise was . . . how naturally skillful all the parents were and the regularity with which we saw optimum conditions for language learning," wrote Hart and Risley. Ultimately, all the kids in the study "learned to talk and to be socially appropriate members of the family . . . with all the basic skills needed for preschool entry."

But while there were broad similarities, the data also showed striking differences. One was observed from the beginning: the number of words spoken in one family versus another.

"After only six months . . . observers could estimate the number of hours of transcription they would need [for each] family and had begun [alternating visits to] a 'heavy' family with one in which there were frequent periods of silence." During each hour-long session, observers found that some families spent more than forty minutes interacting with their child, while others, less than half that.

Cumulatively, the differences were staggering. So was the relationship to socioeconomic status.

In one hour, the highest SES children heard an average of two *thousand* words, while children of welfare families heard about six *hundred*. Differences in parental responses to children were also striking. Highest SES parents responded to their children about 250 times per hour; lowest SES parents responded to their children fewer than 50 times in the same period. But the most significant and most concerning difference? Verbal approval. Children in the highest SES heard about forty expressions of verbal approval per hour. Children in welfare homes, about four.

These ratios remained consistent throughout the study. The amount parents talked to their child during the first eight months of observation was predictive of the amount those parents would be talking to their child at three years of age. In other words, from the beginning until the end of the study, parents who talked continued to talk, and those who did not never increased their verbal interaction with their child, even when the child began to speak.

The data answered the paramount question: Was a child's ultimate ability to learn related to the language heard in the first years of life? Three years of painstaking analysis left no doubt. It did. Counter to prevalent thought at the time, neither socio-economic status, nor race, nor gender, nor birth order was the key component in a child's ability to learn because, even within groups, whether professional or welfare, there was variation in language. The essential factor that determined the future learning trajectory of a child was the early language environment: how much and how a parent talked to a child. Children in homes in which there was a lot of parent talk, no matter the educational or economic status of that home, did better. It was as simple as that.

THE STUDY RESULTS

From Thirteen Months to Thirty-six Months

Children from professional families heard	487 utterances per hour
Children from working-class families heard	301 utterances per hour
Children from welfare families heard	178 utterances per hour

Extrapolating to One Year
The Staggering Difference

Children from professional families heard	11,000,000 words per year
Children from welfare families heard	3,000,000 words per year
The difference	8,000,000 words per year

Thirty Million Words
The Cumulative Difference

Number of Words Heard by the End of Three Years of Age

Children from professional families heard	45,000,000 words
Children from welfare families	13,000,000 words
The difference	32,000,000 words

The Difference in Children's Vocabulary at Three Years of Age

Children from professional families	1,116 words
Children from welfare families	525 words
The difference	591 words

The Real Differences

IQ
Vocabulary
Language Processing Speed
The Ability to Learn
The Ability to Succeed
The Ability to Reach One's Potential

The essential wiring of the human brain, the foundation for all thinking and learning, occurs largely during our first three years of life. We now know, thanks to careful science, that optimum brain development is language dependent. The words we hear, how many we hear, and how they are said are determining factors in its development. The significance of this cannot be overemphasized since this window of time, if neglected, may be lost forever. When Hart and Risley looked at their data, the influence of early language on a child was unmistakable, the negative impact of a poor early language environment critical, including the effect on vocabulary acquisition. Even more significant was evidence of the effect on IQ at three years of age.

"With few exceptions, the more parents talked to their children, the faster the children's vocabularies [grew] and the higher the children's I.Q. test scores at age three and later."

But quantity of words was only one part of the equation. While the number of words a child heard was important, imperatives and prohibitions appeared to stifle a child's ability to acquire language.

"We saw the powerful dampening effects on development when [a child's interaction with a parent] began with a parent-initiated imperative: 'Don't' 'Stop' 'Quit that.'"

Two other factors seemed to have an effect on language acquisition and IQ. The first was the variety of vocabulary the child heard. The less varied the vocabulary, the lower the child's achievement at age three. The other influence was family conversational habits. Hart and Risley found that parents who talked less produced children who also spoke less.

"We saw all the children grow up to talk and behave like their families." Even "after the children learned to talk and had all the skills necessary to talk more than [what they heard at home], they did not; the amount they talked [was identical to what they heard at home]."

Hart and Risley may have had a hunch about the effect of language on learning, but even they were amazed at how well they had predicted the outcome. When they and their colleague Professor Dale Walker reexamined the children six years later, they found that the amount of talk the children had been exposed to through age three also predicted their language skills and school test scores at ages nine and ten.

The importance of the study's finding that the greatest impact on language, school performance, and IQ was not socioeconomic cannot be overstated. Hart and Risley's groundbreaking study showed with statistical force that the preliminary factor in what would eventually become known as the "achievement gap" was the difference in early language exposure. And while at first glance their data seemed to relate this to socioeconomic status, careful analysis linked it solidly to a child's early language experience, which was often, but not always, linked to socioeconomic status.

But perhaps their most important finding was that while the lack of academic achievement in these children was a serious problem, it was potentially correctable by well-designed programs.

CAN WE BELIEVE THEIR RESULTS?

I posed this question to my friend and research colleague Flávio Cunha, PhD, associate professor of economics at Rice University, whose research focuses on the causes and consequences of poverty. In addition to being brilliant, which is the usual assessment of this economist, Flávio Cunha is nice, a wonderful combination of attributes. A protégé of Nobel Prize–winning economist Professor James Heckman, who demonstrated scientifically the tremendous societal cost savings of investing in early childhood, Flávio Cunha provided the following assessment of the Hart and Risley study.

The problem with the study, he believes, is that from a sample of thirty one-hour recordings, Hart and Risley determined a child's entire vocabulary. "It's like my saying that your entire vocabulary is the words you use in this book because those are the only ones I observe." In addition, he continued, while all the recording sessions lasted the same amount of time, because some children spoke less frequently, it would be impossible to accurately know how many more words those children knew. Even more important is being able to gauge the effect of parental speech, since in homes where parents spoke more, children responded more, and in homes where parents didn't speak much, children were less likely to speak much, as well. In that case, recordings may have been more an indication of how parental speech encourages a child to speak, or not, than an assessment of acquired vocabulary.

But two important elements, according to Flávio Cunha, do lend credibility to Hart and Risley's results: their use of established measures of intellectual development, including the Stanford-Binet, and, more important, the substantiating long-term follow-up data. The confirmed impact of early language on school readiness and long-term achievement is a powerful validation of Hart and Risley's study and its conclusions.

But is it really possible to make such strong conclusions from a study that included only forty-two children, each observed for only one hour per month for only two and a half years? Can the thirty-one hours each child was recorded be representative of the fifteen thousand waking hours of that child? And, importantly, can those thirty-one hours actually be predictive of that child's future?

Or is this akin to Mark Twain's "lies, damned lies and statistics?"

The overall goal of the Hart and Risley study was to see if factors in a child's early years could be related to the child's eventual school performance and, if so, whether a well-designed program could improve the child's ultimate academic success. More specifically, Hart and Risley wanted to see if there was something in the early experience of children of higher SES parents that set them on the right track for academic excellence that was missing in the homes of children of the poor.

Initially, there was some concern that the broad interpretation of the data went way beyond their actual scope. As quoted in their article "The Early Catastrophe," "Researchers caution against extrapolating their findings to people and circumstances they did not include." But in the end Hart and Risley agreed that their data demonstrated the predictive power of the early language experience on the eventual academic success of the child, even hinting at programs that might ameliorate the problems.

Betty Hart and Todd Risley may, in fact, have actually underestimated the significance of their findings. By mandating "permanence" and "stability" as requisites for being part of the study, they eliminated that part of society called, by William Julius Wilson, "the truly disadvantaged"; the children, as Shirley Brice Heath described them in 1990, who live "in public housing with single mothers . . . in virtual silence." Had this segment of society been included, Hart and Risley might have found a word gap even greater than thirty million.

Is It Just About Quantity?

Even without science, we know intuitively that saying "shut up" thirty million times is not going to help a child develop into an intelligent, productive, stable adult. Hart and Risley confirmed that. In homes where there was quantity of words, there were also other essential factors including greater richness, complexity, and diversity of language. There was also another, extremely important, characteristic: affirmative feedback. The language these children heard was much more positive and supportive. Recognizing the important confluence of quantity and quality may have prompted Hart and Risley to call their book *Meaningful Differences.*

Hart and Risley's study answered another question: whether families who talked more just naturally used richer language. Their data showed that, in fact, quantity of language drove its quality and that the more parents talked, the richer the vocabulary became. In other words, if parents were encouraged to talk more, the quality of their language would almost inevitably follow, no matter their socioeconomic status. "We don't have to . . . get parents to . . . talk differently to their children," stated Risley. "We just have to help them [talk] more" and the rest will take care of itself.

The importance of quality has been confirmed by Kathy Hirsh-Pasek, professor of psychology at Temple University and Roberta Golinkoff, professor of education at the University of Delaware, whose research has focused on understanding how infants and young children learn language. They, with their colleagues Professors Lauren Adamson and Roger Bakeman, have found that quality of language is important because it increases a child's exposure to a variety of words. It is the key factor in what Professor Hirsh-Pasek calls the "communication foundation" of early language learning. This foundation, which she likens to "conversational duets," consists of three important characteristics of mother-child shared interactions, true regardless of the socioeconomic status of the child's family:

> **Symbol-infused joint attention:** in which both mother and child use meaningful words and gestures as they share an activity.

Communication fluency and connectedness: the flow
of interaction that connects a mother and child.
Routines and rituals: for example, "my turn, now
your turn" play or structured daily events such
as meals or bedtime.

These components of communication work together to create
the optimal context for language learning, says Dr. Hirsh-Pasek,
who emphasizes that her work has been enhanced by the work
of many others in the field.

The Marriage Between Quantity And Quality
The Importance Of Chit-Chat

Hart and Risley's book, *Meaningful Differences,* in addition to
addressing the quantity of words, identifies their functions,
which they label "business talk" and "extra talk." *Business talk*
got "the work of life done" and moved life forward; *extra talk*
was the spontaneous "chit-chat," the icing on the cake.

BUSINESS TALK
"Get down."
"Put your shoes on."
"Finish your dinner."

EXTRA TALK
"What a big tree!"
"This ice cream is yummy."
"Who's Mommy's big boy?"

It was Hart and Risley who were instrumental in giving "extra talk," the chit-chat of life, the attention it deserved. Until then, except for prescient social scientists like Harvard professor Catherine Snow, no one thought twice about the mother babbling to her two-year-old about the juicy red apple going "crunchity-crunchity," while he's eating it, or the off-key chant while she's changing her baby's diaper: "Who's Mommy's sweet, stinky baby?" Yet this was where Hart and Risley found a significant difference in children's early language environments. All children, from every socioeconomic level, had to get the work of life done, that is, had to "sit down," "go to sleep," "eat dinner." But not all children experienced the spontaneous banter, the fun give-and-take that seemed to have an extraordinarily rich effect on the development of the child.

Something else also became apparent. While all types of talk were initiated in relatively equal amounts by all socioeconomic groups, the continuation of talk, the verbal back-and-forth, was where socioeconomic differences also became apparent. Higher socioeconomic status families tended to continue the back-and-forth verbal interaction they had initiated. Lower-income parents, on the other hand, began, then stopped. One utterance, one response, then nothing more. This difference was critical, because contained in the "extra talk" were the essential nutrients for rich brain development. Hart and Risley called this sustained parent-child verbal interaction a "social dance," with each step, or response, increasing in a verbal complexity that further enhanced the child's intellectual development.

To me, however, the most crucial difference was in the use of affirmatives versus prohibitions: "Good job!" versus "Stop it!"

Higher socioeconomic status families reprimanded their children, but far less than parents in the lowest socioeconomic group. Children in poorer families heard more than double the amount of negative remarks per hour than children from professional families. And this difference was compounded by the difference in the total number of words children heard. Because they heard far fewer words in total, the *ratio* of prohibitive, negative words to positive, supportive words was much higher for children in lower socioeconomic status families.

Hart and Risley's study found that children in lower socioeconomic homes were also much less likely than their more affluent counterparts to receive verbal encouragement: "You're right!" "That's good!" "You are so smart!" Children of professionals received about thirty affirmations per hour, about twice what working-class children heard and, painfully, five times more than welfare children heard.

"You're Good/You're Right" Versus "You're Bad/You're Wrong"

In One Year

	Affirmations	Prohibitions
Professional children heard	166,000	26,000
Working-class children heard	62,000	36,000
Welfare children heard	26,000	57,000

Note that the ratio of praise to criticism is reversed in children of welfare parents compared to children of professional families. Hart and Risley extrapolated these numbers to children at four years of age.

"You're Good/You're Right"
Versus
"You're Bad/You're Wrong"

At Four Years Of Age

	Affirmations	Prohibitions
Professional children heard	664,000	104,000
Welfare children heard	104,000	228,000

To better understand this, think how *you* would be affected by hearing one versus the other. What is it like to hear, again and again, that you are wrong, you are bad, you never do anything right? This is a childhood environment hard to overcome, no matter how much your parents, in fact, loved you.

Proof Of This
The Belief Gap

To Shayne Evans, the dynamic CEO of the University of Chicago Charter School, the "belief gap" is a key factor in the lack of achievement in children of poverty, and it's the result of the relentless reconfirmation of inadequacy. If someone tells you, over and over, how little you're worth, especially when it's someone you're expected to believe, how much will you think you're worth? And Shayne Evans says it's what these children hear, not just from their parents, but from school systems, teachers, and society itself.

Shayne Evans's goal for his charter school is to create "a new normal" for these students. In an environment that emphasizes college graduation as an expectation for everyone, regardless of socioeconomic status, family challenges, or any other traditionally inhibiting factor, Evans asserts that "it's our job as educators to help [these students] overcome all obstacles."

Important Points

While the disparity between the highest SES families and welfare families is dramatic, it's important to recognize that the Hart and Risley study showed that disparities exist in a step-by-step gradation starting with the highest SES families, with incremental declines edging down to middle- then lower-income families, then culminating with the startling difference in welfare families. While the difference between professional and middle-income families was not the extreme of thirty million words, it was, nonetheless, a significant twenty million.

It's also important to stress that while we are talking about thirty million words, we are not talking about thirty million *different* words. That would be an extraordinary feat considering that *Webster's Third New International Dictionary* has only 348,000 entries and the latest *Oxford English Dictionary* only 291,000. Rather, what we are talking about is the total number of words spoken, even when those words are repeated.

Hart and Risley were pioneers in a world that could have easily dismissed them. What they became, however, was the first sentence in an important scientific narrative on the impact of early language on a child's life and the critical disparities between children from advantaged backgrounds and those born into poverty. Ultimately, Hart and Risley fulfilled their original goal: to show what was needed in a well-designed intervention, beginning at birth, to help children at risk grow up stable and productive, fulfilling their potentials, and changing the course of their lives.

The Brain And Language Processing Speed

Why didn't the children in the Juniper Gardens Project improve academically? The language processing research of Anne Fernald, professor at Stanford, suggests a profound reason for this. The thirty million word gap, she has shown, is really about the brain and its development.

When Hart and Risley poured vocabulary into the children of Juniper Gardens, they seemed to have found a way of improving their poor school prognosis. At first the project looked very promising; but only at first. Ultimately the children were no different from the other at-risk children entering school. What Hart and Risley did not yet understand, and would not understand until they had completed their study, was that the children, although only four and five years of age, had already been negatively affected by a poor early language environment. While words could be poured into their brains, those words could not improve their ability to learn. Why? Because the effect of that poor early language environment had been on their brain's language processing speed.

A brain's language processing speed, says Professor Fernald, refers to how fast you "get to" a word you already know, that is, how rapidly it becomes familiar and makes sense to you. For example, if I showed you a picture of a bird and a picture of a dog and asked you to look at the bird, how fast would you look at the bird instead of the dog?

It is a process critical to learning. It is, in fact, of double importance, because if you have to work hard at recognizing a word you already know, you also miss recognizing the word following it, making learning exceedingly difficult.

A prime example is conversing in a foreign language that you kind of know. Anne Fernald gives the example of an American student who visits France after having aced her French course, all As, in fact. As she's conversing with someone from Paris whom she's just met, and the conversation is in natural conversational flow, instead of the familiar, easy beat of her professor's speech, she finds she has to "hook onto" each semi-familiar word to "get it," but by the time she's "gotten" it, the conversation has moved on to another sentence. This is, Professor Fernald says, a prime example of "the cost of processing slowly." If you have to concentrate on retrieving the meaning of one word, then all that follows is lost, as well.

While the difficulty in speaking a foreign language has an element of humor, there is nothing funny about a young child's inability to learn. What Anne Fernald found when she studied toddlers in a lab was that a split-second delay in grasping the meaning of a familiar word in a sentence made it much more difficult for a child to figure out the next one. A simple hundred-millisecond advantage, she says, "buys you the opportunity to learn." For those without that advantage, the loss can be incalculable and permanent.

Anne Fernald found the same socioeconomic relationship

that Hart and Risley had. In her study, two-year-old children from low-income backgrounds had a full six-month lag in vocabulary and speech processing skills compared to children from higher socioeconomic status.

But Professor Fernald also confirmed that while socioeconomic differences were apparent in the data, those differences were not the salient ones in the results. Her study of children from *only* low-income backgrounds found a huge variation in how much parents talked: with a range of 670 words per day to 12,000 words per day. It also found a significant relationship between a child's early language environment and the child's language processing speed without the factor of socioeconomic status. At two years of age, children who had heard less talk had smaller vocabularies and slower language processing speeds. Those exposed to more talk had larger vocabularies and faster language processing speeds. And that was true for all socioeconomic levels.

It all came down to how well the brain had been nourished with words.

CHAPTER 3

NEUROPLASTICITY

RIDING THE REVOLUTIONARY WAVE
IN BRAIN SCIENCE

Biology gives you a brain. Life turns it into a mind.
—Jeffrey Eugenides, *Middlesex*

A large percentage of our physical brain growth is complete by the time we're four years old. The ease with which we learn as children and the design of our entire lives are heavily predicated on what happens in those first years. Why is this a painful truth? Because it comes at a time when babies have no power to say, "Whoa, you're doing this wrong!" "Talk to me more!" "Please, speak nicely to me." Just as a baby who is starved of adequate food in the first three years may survive but will never grow to his or her potential height, a baby whose brain is starved of adequate language will survive but will have enormous difficulty learning and will never reach his or her full intellectual capacity.

Science corroborates this. Anne Fernald's elegant study demonstrates that the language processing speed of a child whose early language environment is poor is slower and less efficient. Hart and Risley saw the effects of this as well when the preschool children in their project, after receiving an intensive vocabulary intervention, showed no difference in their ability to learn. Their intervention was powerful, yet it still could not improve brains impaired by a poor early language environment.

To understand why, we have to know how the brain, our most extraordinary organ, develops and why the early language environment is the catalyst for who we are and what we can become.

A BABY'S BRAIN
A WORK IN PROGRESS

The brain, unlike almost all other organs, is unfinished at birth. The heart, kidneys, and lungs function from day one as they will for their entire lives. But the brain is almost entirely dependent on what it encounters on its ride to full development. Inside that cute, lovable newborn is an intellectual core on the brink of incredibly rapid, complex, intricate growth.

Within a few years after birth, a relatively small blip of time, a brain circuitry that is remarkably strong or dangerously weak or somewhere in between will be created, affecting a lifetime of attainment. What are the key factors determining this? Essentially genetics, early experience, and their lifelong effects on each other. That's it; for better or worse.

Dr. Jack Shonkoff, the director of Harvard's Center on the Developing Child, compares the development of a baby's brain to building a house. "*Genetics,*" says Dr. Shonkoff, "supplies a basic plan for brain development . . . just as an architect supplies a blueprint for building a house. The genetic plan . . . lays down the basic rules for interconnecting nerve cells . . . [providing] the initial construction plan for the brain's architecture," ultimately determining what is developmentally unique to each of us. While this genetic potential determines the "ceiling" in various spheres for each of us, for example, I will never be as smart in economics as Nobel laureate James Heckman, no matter what my early experience was like, there is a huge discrepancy between reaching the heights of one's various potentials and being trapped at the bottom in all possible spheres.

What Jack Shonkoff is saying is that in the building of a house even the grandest of blueprints will not compensate for the lack of quality materials, skilled contractors, or a crew of supportive workers. Without those, the finished product will never be what the architect had envisioned, nor will the house ever be what it might have been. This translates easily to the young child. The one thing that all babies have in common is complete dependence on others for absolutely everything. Traditionally, this has been recognized as the need for nourishment, that is, milk, to survive and to grow. Only now are we beginning to understand that, in addition to the food necessary for physical growth, there is an equivalent need for optimum social nutrition to ensure intellectual growth. And both needs are absolutely caregiver dependent.

An important part of that social nutrition, now recognized

as essential to achieving optimum brain development, is stability. The developing brain is hypersensitive to all stimuli in its environment. A "toxic" environment, filled with constantly high levels of stress during infancy, has been shown to produce internal "stressors" in babies. These stressors represent the first inhibiting factors in a baby's brain development, demanding so much of the brain's attention that the brain is diverted from learning. Some stress is, of course, a part of all lives, even babies', a delayed feeding or crying before sleeping, for example. But when the levels of stress are consistent and consistently high, "stress hormones" such as cortisol bathe a baby's or young child's brain, forever changing its architecture. The results of brains permanently altered by stress include chronic behavior problems, health issues, and learning difficulties.

The irony is that children who have been reared in environments without chronic stress are better able to deal with life's "turbulence," that is, stress, in more constructive, less negatively reactive ways.

The most important component in brain development is the relationship between the baby and his or her caretaker, which includes the ambiance of the language environment. Who would have guessed that cooing "Daddy loves his honey bunny" to an infant who is just beginning to focus could be that important? But it is; very, very important, as a matter of fact. In tiny step-by-step increments, the "oohs" and "aahs," the "Mommy loves you," and the "What a sweetie pies" are the catalysts quietly connecting the brain's billions of neurons to create the complex neural circuitry that will culminate in a child's intellectual potential being realized. When the scenario is optimum, with cooing and smiling and peacefulness, the brain develops beautifully. When these optimal conditions do not exist or, even worse, when the environment is shrill and isolating, brain development is severely and negatively impacted.

In the end, quantity of words is important, but only as an adjunct to the loving, nurturing relationship that is determined by a baby's caretaker. There may be many words, but their positive effect on the brain is dependent on responsiveness and warmth.

THE STILL FACE EXPERIMENT

For the development of vision, the environmental catalyst is easily understood: the light of day. For the developing mind, the environmental spark is much more nuanced and far more complex. It's in the returned gaze of a mother to her baby, a father picking up his child as she holds her arms up to him, a parent saying "juice" while handing a cup to the child, or playing peekaboo to elicit a smile or a giggle. This positive, responsive back-and-forth, serve-and-return, is the foundation on which a lifetime of learning, behavior, and health is built. At its heart, in fact at the heartbeat of brain development, is a baby's relationship with a caring, responsive adult.

A very touching example of a baby's need for social interaction exists in the unforgettable "still face" experiment on YouTube done by University of Massachusetts distinguished professor of psychology Edward Tronick.

In the video, a young mother buckles her baby into a high chair and proceeds to play with her. Suddenly the mother turns her head away from her child. When she turns back, her face is blank, expressionless. The baby stares at her, confused; then suddenly this baby, whose joy had radiated like sunshine, begins frenetically pointing, reaching out, trying everything to elicit a response from her mother. But to no avail; her mother continues to stare blankly at her. The baby, realizing the futility, arches backward and begins wailing. It is painful beyond belief to watch.

But the baby is not the only one affected. Now we see the mom begin to exhibit anxiety until finally she can take it no longer. Her blank face transforms into the loving mother she is and her baby is almost instantly happy again.

In real life, loving mothers rarely play this game. But it's also true that for some babies, it's more than a game; it's their lives. Living in environments that are chronically "blank faced" or, even worse, angry and hostile, is not something that, in a few seconds, will be rectified with a hug. In those cases, as we've discussed, stress hormones such as cortisol begin bathing the babies' brains, the profoundly negative, often irreversible, effects permeating their core. The result is observable not only in cognitive and linguistic development, but also in behavior, self-control, emotional stability, social development, and overall mental and physical health.

This reinforces the sad truth that a child's genetics, its blueprint for potential, the gifts handed down by birth, are not, in fact, set in stone. Epigenetics, the process by which genetics are altered by environmental influences, demonstrates that nurture may not be able to improve nature, but it can harm it. Early "toxic" experiences, including highly stressed environments, have been shown to be able to change the genetic blueprint profoundly and negatively, permanently affecting the developing brain. It's important to emphasize that we're talking about constant, chronic, relentless stress, not the occasionally disgruntled, very tired mom or dad "It's 2 AM, baby, please, please, give me a break! All right! All right! I'm coming!"

THE MAGIC OF A FUNCTIONING BRAIN

Each of us is born with the potential of one hundred billion neurons. That translates into a lot of potential. Unfortunately, without the critical neural connections, those one hundred billion neurons are relatively meaningless, rather like freestanding telephone poles with no connecting wires. When, on the other hand, those neurons are optimally connected, it is their speedway-fast signals to one another that allow the brain to do its magic.

From birth through about three years of age, each second represents the creation, by the brain, of seven hundred to one thousand additional neuronal connections. Let me run that number by you again: *seven hundred to one thousand additional neuronal connections every second of a baby's life*. The incredible, complex circuitry that results is the brain's architecture, affecting all our brain function, including memory, emotion, behavior, motor skills, and, of course, language.

But this explosive profusion of neural connections in the first three years turns out, in fact, to be way too much. If it were allowed to remain, the brain would be a chaotic overload of stimuli and noise. So, through a process called synaptic pruning, our very smart young brains begin to trim away superfluous neural connections, weeding out those that are weaker or less used and fine-tuning those used often into areas of specific specialized functions.

It is during this period, when the critical neural connections are being created and strengthened, that the potential for skill building and verbal learning is extraordinary. Never again will the brain have the same degree of neuroplasticity, that is, the remarkable flexibility to change in response to different environments. But as this window of opportunity narrows, as the brain begins to prune unused or little used connections, the potential scope of adaptability to a wide range of opportunities also narrows, making new endeavors, for example, learning a new language as you get older, increasingly difficult. This is a time, Dr. Jack Shonkoff says, "of [both] great opportunity *and* vulnerability."

Abdullah

Abdullah, a twenty-year-old deaf student at a local community college, had come to see me about a cochlear implant. The seven of us, Abdullah, his Palestinian immigrant parents, his little brother Mohammad, two interpreters, one signing and the other speaking Arabic, and I, crowded into the clinic room. Mohammad, the only one in the room not using an interpreter, ran interference, switching between English, Arabic, and sign language with ease. Just nine, with big brown eyes and a layer of baby fat, Mohammad had the confidence that undoubtedly came from being the voice for his parents and a much older brother. Fluent

in English and Arabic and familiar with sign language, Mohammad was a portrait of optimal neuroplasticity. He and his family
had come because Abdullah had recently decided that he wanted
a cochlear implant, to "hear and connect," he told me via his
interpreters.

"Realistic expectations," our field's code for discussions of
the potential for success in cochlear implantation, is an essential
part of my counseling with older patients like Abdullah. It is
based on the degrees of neuroplasticity, the brain's ability to
form new neural connections for learning, or the lack thereof.
For Abdullah, because of his age, it was likely he would never
speak, or understand spoken language, or do the things normally associated with being able to hear. Sign language would
almost undoubtedly continue to be his mode of communication.
His "realistic expectation" would most likely be sound detection: an airplane overhead, a doorbell ring, a toilet flush, rain
hitting the windowpane. But hearing these sounds and understanding their meanings were two different things. His brain, I
had to explain to the hopeful family, had passed the *critical
language period.*

His parents listened politely, as did Abdullah and his brother.
Finally his mother said, via the Arabic interpreter, "Doctor, I just
want for my son to be helped." But her eyes, framed by her hijab,
told me she was hoping for much more. My explanation, my
description of "realistic expectations," were no match for that
hope. If her first-born son were given the ability to hear, why
wouldn't he then, automatically, understand what he was hearing? Or be able to speak? This time I spoke to her, one mother to
another. It was, I explained, like my moving to Palestine expect-

ing to understand the Arabic words that surrounded me just because I could hear them. "Mohammad," I said, "might just have to come and interpret for me." She looked at Mohammad and a wistful smile spread across her face. Now she understood.

Abdullah, a lovely young man, bright, supported by a wonderful family, unlike my very young patients, no longer had the neuroplasticity, the potential for understanding the sounds that he would now be able to hear.

It was all in the timing.

CRITICAL PERIODS
IN
BRAIN DEVELOPMENT

The visual system is one of the best-studied areas of humans. The image we perceive when we look at something, including its form, color, detail, and depth, is the brain's reconstruction of the image captured by our eyes' retinas. And, like most brain functions, vision is also a skill that becomes fully developed only after birth. During the first few months of life, babies can see a distance of only eight to ten inches and can barely coordinate their eyes. Within a few months, however, coordination dramatically improves, until gradually, over the next two years, depth, color perception, and visual appreciation of the world explode.

Like language, however, visual acuity is also environmentally dependent. Simply put, a baby needs things to look at in order to be able to see.

What happens, then, when the visual environment is not

there, for example, when a baby is born with a milky-white film covering his or her eyes during the visual system's "critical period?" Exactly what happens with other brain functions: The brain goes into its "use it or lose it" mode and begins its pruning process, weeding away unused or weak neural connections, in this case the barely stimulated visual receptors. The result is a child who will probably never see well, even if the film is ultimately removed.

This became evident in the early 1900s, when eye surgeons found that operating on babies born with cataracts restored vision completely, with no lifelong consequences. When, however, they operated on children with sight-impairing cataracts who were more than eight years old, the children's eyes would look normal, but their visual problems continued throughout their lives. It's a problem similar to the timing of cochlear implantation.

The essential question was, of course, why? Torsten Wiesel and David Hubel's explanation revolutionized our understanding of brain plasticity. Their Nobel Prize, awarded in 1981, was for discovering "one of the most well-guarded secrets of the brain."

Hubel and Wiesel began their research to gauge the activity of individual neurons in cats and monkeys in the 1950s. In addition to conceptualizing the study, they had to create new instruments to measure the animals' brain responses to what they were seeing. Clever as well as innovative, their array of novel approaches included placing research cats "adorned in electrical headgear, before a screen that displayed all sorts of visual images . . . trying to find a stimulus that could coax a single neuron . . . to fire." The images, or so the story goes, included the two of them dancing and sexy pictures of women. "When it comes to sheer fun," Hubel wrote, "our field is hard to beat. We try to keep that a secret."

Although they worked with vision, Hubel and Wiesel's seminal study changed the way we understand the brain. Nobel Prize–winning neuroscientist Eric Kandel described the importance of their work beautifully and succinctly. When a fellow scientist declared Hubel and Wiesel's work as having "limited biological generality," Kandel responded, "You're right . . . It only helps to explain the workings of the *mind*."

IT'S ALL IN THE TIMING

Just as you can't run before you can walk, you can't say your first word until you hear and understand that word. The repercussions from missing the timing for a skill are severe because, in brain development, acquiring a basic ability is a prerequisite for acquiring a more complex one, each skill acting as a building block for the next. Brain development, in other words, occurs in a hierarchical fashion, with the "basic" abilities providing the foundation on which the more complex ones are built. Missing the window for a "simple" skill, therefore, has wide-ranging implications because, while new learning may occur, it just becomes more and more difficult. This is especially critical in language accrual, because language, during the first three years, in addition to helping build vocabulary and conversational skills, helps provide a foundation for social, emotional, and cognitive development.

Clear examples of an inadequate early language environment are adults who were born deaf to parents who, although loving, did not have the ability to sign. The lives of those adults often demonstrate the consequences of a very real early word gap.

My Second Cousin

My mother's cousin was born in 1948 with a profound hearing loss. As a child, I vaguely remember receiving long, rambling, handwritten birthday greetings, cards I barely glanced at. If there wasn't a gift attached, cards meant very little to a nine-year-old. It was only recently, when my mother was reading an early draft of this book, that she casually mentioned that her aunt and uncle, a schoolteacher, had, in fact, moved from Pittsburgh to St. Louis just so their only child could go to the St. Louis Central Institute for the Deaf, a school committed to "spoken language" rather than signing. A quick search of Central Institute's archives found him. I suddenly realized that my distant cousin with, incredibly, a blue and a brown eye, the result of Waardenburg syndrome, was almost identical to my patient Michelle with the brilliant sapphire blue eyes. But unlike Michelle, my cousin had been born to parents who had the financial means, and the will, to move across the country for their son's education. In grand irony, he had gone to the school that was down the street from where, almost forty years later, I would perform my first cochlear implant.

What had happened to him? What had his life been like?

While she would not go into detail, my mother said that her cousin's life had not been easy. She didn't know his literacy level, although until a few years ago they corresponded regularly. If he were similar to the average deaf child who had been born to hearing parents during that time, however, before the advent of cochlear implantation and access to hearing, his ultimate literacy level, no matter the advantages of parenting, would likely have been at about the fourth-grade level. While this was typical for the times, it did not necessarily reflect his potential at birth. Quite the contrary, there is a strong probability that his potential could never have been met because he was unable to hear. He was, as his peers often were, a victim of the thirty million word gap in the purest sense.

My cousin's experience is evidence that it is not the socioeconomic strata we're born into, nor even the intent of our parents, that is the salient factor in our growth as humans. If that were the case, my mother's cousin's ride through life would have been a smooth one. What he lacked was the nutrition of "words," which, when it is absent in a young child's life, either through signing or speech, has a permanent, negative impact, no matter whom we're born to.

It's important to stress that it was not only my cousin's life that had been affected. While the cochlear implant has brought both sound and the possibility of fulfillment of potential for all the Zachs in the world, society is also a major benefactor. Given the costs of special education, underemployment, and unemployment, deafness is one of the most expensive of disabilities. The cochlear implant is an essential key to avoiding that cost but, as we can see from Michelle's story, a key is of little importance unless it is used to open a door.

HEARING AND READING
AND
LEARNING

Learning to read, for those who hear, seems a relatively easy, sequential process of learning letters, learning their sounds, learning their combinations for words, and learning what those words mean. For the deaf, on the other hand, reading is an extraordinary challenge. In fact, "challenge" is a euphemism; it is a Herculean task.

Imagine, if you read only English, having to learn words you don't know, written in Chinese characters. In the same way, deaf children are being asked to recognize letters on a page, combine them for words, and understand the meanings of those words without ever having heard them. The word "cat," for example; easy, right? You know the "ka" sound of C, the "a" sound of the A, and the "t" sound of the T. And you immediately equate the combination of those sounds to a little furry animal that says "meow."

But what if you'd never heard the sounds of the letters C, A, T, either individually or strung together? What would those symbols mean to you? Even though you live in a country where the word "cat" is universally known, even though you can sign for the animal "cat," seeing C-A-T means nothing. That is the arduous road that a deaf child has to go through to learn to read. Knowing sign language does not help because sign language consists of motions that indicate meaning but are not English in its written form. Sign and English are, in fact, two different languages; there is absolutely no connection. As a result, when young deaf children learn to read, they are in constant translation mode from sign language into English, without ever having actually heard English or knowing how it sounds. Herculean may, in fact, be an understatement.

The consequences of this are enormous. Hearing children begin school by learning to read with the ultimate goal of reading to learn. Third grade is the pivotal year when children flow from simply sounding out words on the page to forming ideas and accumulating knowledge from those words. It is the beginning of intellectual thought processing, but only for those who can read proficiently. For those who can't, third grade is also significant, because it's the beginning of a documented sharp decline in knowledge accumulation and intellectual growth.

Psychologist Keith Stanovich calls this the "Matthew Effect," after the Gospel of Matthew: "For whosoever hath, to him shall be given, and he shall have more abundance: but whosoever hath not, from him shall be taken away even that he hath" (Matthew 13:12). In other words, the educationally rich get richer and the educationally poor get poorer. Third-grade reading is so influential, in fact, that it predicts high school graduation. And this is where the significance of deafness is profoundly apparent. The potential for a deaf child's graduation from high school or college is significantly less than for a hearing child. The resulting impact on employment is undeniable. Historically, underemployment among the deaf is tragically common, and those who *do* work earn 30 percent to 45 percent less than their hearing counterparts. When you read these statistics, it's very important to remember that we are not discussing differences in intellectual potential. We are describing individuals whose potential we will never know.

On The Other Hand

When the language environment is optimum, however, there's an entirely different ending. Like other facets of brain development, language acquisition follows the skills-begetting-skills path, with each learned proficiency laying the foundation for the next. It seems to happen so automatically that we generally take it for granted. In fact, a newborn's going from hearing the continuous stream-of-gibberish sound:

"Whosmommyssweetiepie?"
to hearing each word discretely:
  "Who's Mommy's sweetie pie?"
then understanding that each segment has meaning:
  "Who's"
  "Mommy's"
  "Sweetie"
  "Pie"

. . . then being able to say those sounds themselves and, finally, to even answer the question, remains an amazing, almost unfathomable, feat of human development. No matter what language a child is born into, whether in rural Tanzania or metropolitan Manhattan, the developmental paths are essentially the same, with language input, quantity, and quality, as the key catalyst for the developing brain.

"Who's Mommy's sweetie pie?" (English)

"Kas yra mamytė savo saldainiukas?" (Lithuanian)

"Aki a mama a kicsim?" (Hungarian)

"Thì pĕn fæn k̄hxng mæ̀? (Thai phonetically)

"Ambaye ni mama ya sweetie?" (Swahili)

Think of hearing a sentence in a language you don't speak. Gibberish, right? So how is it that a tiny baby, encountering a river of sounds totally alien to its ears, transforms this babbling stream of sound into segments of sound . . . *phonemes* . . . then translates those meaningless segments of sound into words that ultimately convey meaning? It's an incredible journey that neuroscience has only recently begun to explain.

Professor Patricia Kuhl is a pioneer in our understanding of how babies crack the speech code. I first learned about her innovative studies during the Introduction to Child Language Development course with Susan Goldin-Meadow. Using techniques as simple as monitoring a baby's rate of pacifier sucking when it hears a sound, Professor Kuhl has painstakingly uncovered the steps by which babies learn language. In addition, her sophisticated new tool, a magnetoencephalograph, or MEG, which she has described as "a hair dryer from Mars," gives her real-time images of a baby's brain in action or, as she says, "a peek under the hood" of a child's brain. What she has found in her research, says Professor Kuhl, is that babies really are, in fact, "computational geniuses."

WE WERE ALL
COMPUTATIONAL GENIUSES

Before any of us understood, or said, a single word, our brains had to go into action "parsing," that is, separating, sounds, then splicing them together to create words. This is an important part of the early brain's job of learning the native language, with some indication that the process may actually begin in utero. With the agility of kung fu masters, the remarkable brains of babies slickly slice and dice the incoming stream of sounds until they transform into words with meanings, settling into the context of the language they're part of.

A wonderful anecdote shows that even adult geniuses can't compete with the genius of a newborn baby. Mark Zuckerberg, founder of Facebook, who learned Mandarin so he could talk to his Chinese in-laws, had a thirty-minute meeting with Chinese leaders. Their final assessment of this brilliant Internet entrepreneur's Chinese language skills? That of "an articulate seven-year-old with a mouth full of marbles" who had said that Facebook had a total of eleven users . . . rather than one billion!

In truth, adults attempting to learn a language are no match for babies. Imaging of babies' brains shows that even before they say their first words, they are mentally practicing responding, trying to figure out how to create the motor movements necessary to articulate the words of their language.

WHY CAN'T WE DO IT?

An infant's brain, at the height of neuroplasticity, can distinguish the sound of every language, from the German umlaut to the Chinese pinyin to the glottal, slightly implosive consonants of the Masai, and is ready to learn the language a sound belongs to, or even several languages with very different sounds. Babies are true "citizens of the world," as Kuhl called them. But it is not a skill that lasts forever. Similar to the brain's eventual pruning of synapses that are not used or are underused, the unlimited potential for hearing and uttering every possible sound from any language begins being trimmed away very early, giving us heightened ability to utilize our own language but preventing easy access to the sounds of those we don't use.

That commitment to the sounds of a native tongue occurs very early in a young child's life, usually by the end of the first year. How does the brain, which has been primed to learn the native tongue as early as the third trimester of pregnancy, know which neural connections are to be permanent? Statistical brilliance. A baby's developing brain, incredibly, begins quantifying the specific sound patterns it hears from the beginning, computing their frequency, without any concern for word meaning. The sounds that predominate are retained by the brain, eventually becoming individual words and ultimately becoming the native language.

This is done, in a sense, by a baby's brain "harvesting" repeated sounds and tagging them as "prototypes," important sounds to retain. A prototype sound then becomes, as Patricia

Kuhl has said, like a magnet for similar sounds, even those with slight variations. This process eases us into familiarity with the language we'll be using, but it also hinders our ability to accurately hear or speak languages with dissimilar sounds. Think of the difficulty for those who speak Asian languages to separate the "r" sound from the "l" sound and, conversely, European-language speakers' inability to replicate Asian tones. But this is yet another part of the brain's brilliance because, aware of the need for language, yet also of its own limitations, the brain homes in on the essentials and eliminates the extraneous. After all, why waste precious mental processing for meaningless variations that have no importance in the language you have to speak well?

Patricia Kuhl's early research experience with babies in Japan exemplifies this. At seven months of age, when they were still "citizens of the world," the babies could differentiate between the English "r" and "l" with absolutely no difficulty. On her return three months later, the ability had disappeared. The same thing happened with the American babies she studied, using other sounds. In both cases, the brain, aware of its imminently decreasing neuroplasticity, "commits" to the sounds of the language it will need and refuses to expend neurons on those it will not.

DON'T TRASH BABY TALK

"I never baby talk to *my* baby," brags Mommy. That statement seems to be a universal badge of honor in new parenting, as if baby talk were really bad. Well, surprise, surprise, it turns out baby talk is good. The almost instinctive sweep of "Mummmmy loooooves her litttttle baaaaaby," combining high-pitched speech with slightly altered words stretched out into a singsong stream of sound, is, science shows, a way to help babies' brains parse out sound and commit to the language they'll be using. While it may simply sound like an expression of motherly love, "baby talk" actually helps the baby statistician's brain to more easily grasp sounds that are clearly distinct from others, each acoustically "exaggerated" when compared to adult-directed speech, making them easier to grasp and to learn.

HOW ABOUT TV?

If babies are such computational wizards, why can't we just sit them in front of a television and call it a day? Then we could at least finish the book we're reading. Or answer some emails.

The brain may be brilliant, but unfortunately for that growing list of unanswered emails, it's a social creature. Taking away the interaction may also critically limit its ability to learn and to retain knowledge. Unlike pitchers that will hold anything you pour into them, the brain appears to be more like a sieve without the human interaction. What *is* language, after all? If we lived in a world of isolation, we wouldn't need it, right? The basis of language, of words, is to connect humans to other humans. A baby's brain is a result of that evolutionary history. It does not learn language passively, but only in an environment of social responsiveness and social interaction. The importance of the linguistic serve-and-return in the baby-caretaker relationship is a key factor in learning language and in learning; its importance cannot be emphasized enough.

One of my favorite studies by Dr. Patricia Kuhl demonstrates this point beautifully. Professor Kuhl's team exposed nine-month-old American babies to Mandarin. Half of the babies heard it via mother figures who mixed language with cuddly mother warmth. The other babies heard exactly the same Mandarin, also spoken with mother warmth, but via an audio recording or video device. After twelve laboratory visits, the babies who heard the language via live individuals were able to identify the sounds of Mandarin. Those who heard the same words via a recording or video device? You guessed it. Nothing.

This raises an interesting question. Does it mean that babies learn only from someone they can smell, touch, or feel? Or is it possible that a robot, think Steven Spielberg's *A.I.*, can substitute for this human component? What are the human factors necessary for optimum brain development? These questions are among the millions still to be answered about an incredible organ that has the most profound influence on each of us and on the world in which we live.

NEW HOPE

A child's ease of absorbing new knowledge becomes increasingly difficult as the plasticity of its brain, allowing efficiency and specialization, decreases and the window for effortless learning begins to close. But what if it didn't have to be so? What if that window could be pried open and the brilliant ability of the child to learn became a lifelong phenomenon? Think of learning a new language with relative ease at forty or fifty. Described as brain "time travel," this hypothesis is part of recent studies on the brain geared to understanding it better.

The research of Takao Hensch, professor of molecular and cellular biology and professor of neurology at Harvard Medical School, was inspired by Hubel and Wiesel's studies on brain plasticity. But Professor Hensch has something that Hubel and Wiesel could only have dreamed of: molecular tools that help scientists understand brain responses at the cellular level. With them, he has uncovered another startling new finding: that counter to what was previously thought, the brain does *not* lose plasticity; it appears to have, in fact, the capability of limitless rewiring. Why doesn't this work in real life? Because evolution, for reasons yet unknown, stopped this ability by creating molecules that put "brakes" on, preventing constant rewiring and putting an end date on the plasticity of the brain.

Professor Hensch's pivotal research, done with colleagues at Boston Children's Hospital, involves attempting to reverse molecular braking in order to restore sight in patients with amblyopia, or decreased vision in one eye related to the brain's early neural pruning. While the study is still ongoing, early results appear promising. His study of "tone deaf" men has already shown that when molecular brakes are reversed, ears can be retrained to hear individual musical notes, an ability that, if not cultivated, is supposed to be lost in early childhood.

"What's so interesting about Takao Hensch's work is that he has shown that even if you miss these critical periods, you still may be able to go back in and fix things," says Professor Charles Nelson, a neuroscientist at Boston Children's Hospital, in "Neurodevelopment: Unlocking the Brain." "The idea that you could intervene later and make up for lost time is compelling."

Actually, to me, much more than compelling. While the brain is still an intriguing frontier, the strong indication that someday its mysteries will be uncovered, and that we'll have the lifelong ability to learn and grow, gives me hope that it will also allow us to understand ourselves better, another step toward creating a more humane, just world.

CHAPTER 4

THE POWER OF
PARENT TALK

FROM LANGUAGE TO AN OUTLOOK ON LIFE

I am a brain, Watson. The rest of me is a mere appendix.

—Arthur Conan Doyle,
The Adventure of the Mazarin Stone

In their classic lyrics of 1930, Buddy DeSylva and Lew Brown wrote ". . . the best things in life are free."

Just think about it.

The incredible power that helps nurture the brain into optimum intelligence and stability is *parent talk*. If the most profound mysteries of the brain are still to be discovered, that truth has already been revealed. And it shows you how smart the brain really is, because, in absolute evolutionary brilliance, it harnesses a plentiful, natural resource as the key catalyst for its own development. The process is so simple and hidden that you aren't even aware it's happening. You can't sell it, you can't store it, you can't list it on the New York Stock Exchange, but a caregiver's language is the essential resource of every country, every culture, every person, extending into every crevice of who we are, what we can do, and how we behave.

And it doesn't cost a cent.

THE CONNECTOME

Neuroscience is like an intriguing mystery novel with sharp-brained detectives, who happen to have PhDs, ferreting out clues that will finally let us know, on the last page, why we are the way we are. Of course, the difference between neuroscience and Sherlock Holmes is that we know, from page one, who the perpetrator is: the brain. What these detectives with degrees are trying to discover is how it works, because once they do, once they find out how it makes us who we are, we can help it make us who we want to be.

The importance of the brain has long been acknowledged but, until recently, our understanding of how it worked was rather simplistic and largely based on inference. For example, if a patient diagnosed with a stroke in the left temporal lobe of the brain had lost the ability to understand language, or a patient with a cerebellar brain tumor could no longer swing his golf clubs, physicians attributed the loss to that specific region of the brain. That's it. It was neuroscience in the dark.

Then came the magic of brain imaging, the power of computer science and mathematical modeling and, almost immediately, the superficial appreciation of this amazing organ suddenly changed into a minute, cellular understanding of how it works, not completely, but enough to set us on a path toward finding out completely.

A street map of New York and the wiring of the brain are very similar. Think of Manhattan with its streets and more streets extensively crisscrossing, alive with movement and activities, but still highly organized. Then think of the wiring of the developed brain with its neurons, the specialized cells that transmit information throughout our bodies, and more neurons, one hundred billion of them, in fact, entirely interconnected. That interconnected wiring is called the *connectome*. And that *connectome,* the ten thousand connections *per neuron* linking the one hundred *billion* neurons in each of our brains, is what makes us who we are, including how we think and how we behave.

PARENT LANGUAGE
AND
BRAIN WIRING

"Intelligence" is an awe-inspiring and sometimes daunting word. Intelligent is what we all want to be. How nice to hear, "He's sooooo smart!" "The head on that girl . . . !" It's the facet that, for just about everyone, seems absolutely essential to self-esteem. Not only does it matter how smart people think *we* are, but let's face it, when our kids seem smart to others, we suspect we do, too! Where does intelligence come from? Well, although we all have intellectual potentials in a myriad of spheres, achieving those potentials is quite another thing.

As we discussed in the last chapter, when parents coo to a child, "Who's a cutie-pie?," "Who's the most wonderful baby in the world?," traditionally it seemed like a sweet, not really important part of being a parent. But, quite the contrary. A cheer for those "cutie-pies" and their close cousins, because the connectome, the ever-evolving network of the brain, where neuronal wiring and neuronal pruning make us who we are, is absolutely related to them.

How do we know that?

Mapping the connectome labyrinth in order to discover the complex mysteries of the brain and, therefore, how we become who we are, is definitely neuroscience's new frontier. Until now, questions that have plagued philosophers from the beginning of time could not be answered because the only way of exploring them was through words, debates, hypothesizing, conjecturing. Even now, although new techniques are helping us understand what has always seemed to be the incomprehensible, the answers they reveal often only open up new questions. What is clear, however, is that this vast array of interconnectedness, which is *you,* is "nature meeting nurture." And, while every aspect of the connectome is not yet known, we do know that life's experiences, especially those from birth through three years of age, can profoundly change your connectome and, in the process, change you. This is true even in identical twins, each of whom will have a unique connectome, different from the other.

Similar to the lively streets of Manhattan, each of which has its own purpose but as a whole creates the dynamic, complex metropolis of New York City, while each neuron connection in our brain has its own purpose, the complex network of these connections, that is, the connectome, determines who we are as a whole, including being an essential factor in the achievement of our personal strengths, whether they be in scientific research, writing poetry, or plotting out a winning basketball strategy. Where do those important neuronal connections start? While the aspect of genetics cannot be discounted, achievement of innate potentials, science strongly indicates, is largely determined by a child's earliest language environment, that is, by parent language.

The term "parent language" is misleading, however, since its magic works way beyond simply introducing vocabulary. Referring to the *number* of words a parent speaks to a child and the *way* in which a parent speaks to a child, parent language influences our ability to reach our potentials in math, spatial reasoning, and literacy, our ability to regulate our behavior, our reaction to stress, our perseverance and even our moral fiber. It is also an essential catalyst in determining the strength and permanence of certain neuronal wirings and the pruning away of others.

All of us come into this world with specific strengths and areas in which we are unlikely to succeed. Even the best language environment will not eliminate our weaknesses nor propel us to the highest level of achievement in every endeavor. What science tells us, however, is that reaching the potentials we *do* have is strongly determined by what happens to us when our brains are being developed, from birth through about three years of age. In short, the genetic potentials we've been awarded by the chance of parentage can be mitigated, destroyed, or achieved by our second round of luck, the parental language environment we experience as children. This is something, I believe strongly, that all parents, in fact, everyone, should know.

"I HATE MATH!"

So announced our oldest child, Genevieve, when she was about eleven years old. Math, she said repeatedly and heatedly, was "not [her] thing." Four years and nine inches later, she's a math whiz. In fact, if you asked me now what the pivotal strengths of my daughters and son are, I'd say math. But while my daughter's math ability might elicit an admiring smile, "a girl doing well in math? wow!," for my son, it's often what's expected, just as my daughters' expertise in the humanities, debate, and writing is not surprising to anyone. They're girls, after all.

Now a confession. Early on my husband and I had also, very subtly, succumbed to this preconception. When they were very young, we had joked that our daughter emerged into the world talking in long sentences and our son in long division. At that time, writing about parent talk as a catalyst for a child's math ability would have seemed a bit of a stretch.

We admit it . . . we were wrong! Sorry, Genevieve!

Finding out that we were wrong, and doing something about it, may well have turned the tide for our daughter and her math skills. Our nation's finding out it's wrong, and wanting to correct the problem, may well turn the academic tide for a large population of other children, girls and boys.

The United States recognizes its lagging achievement in math and the need to address the problem. Math's poor standing, often lumped with science, technology, and engineering education (STEM), is becoming more and more apparent as the United States slips precipitously in relation to other developed countries, including, of course, China. The problem is not only for our children and their education, but for the future of this country, productively and competitively.

Elizabeth Green's *New York Times* article "Why Do Americans Stink at Math?" is funny and not so funny. In it, she recounts how, in the 1980s, A. Alfred Taubman, owner of the A&W chain, attempted to win over customers from McDonald's. To lure customers from McDonald's Quarter Pounder hamburger, he advertised the A&W *better-tasting burger* that was, in contrast to the McDonald's Quarter Pounder, a full one-third pounder. One-third of a pound versus one-quarter of a pound and at the same price!

Great idea, right?

Well, not if you don't know one-third is more than one-fourth!

Taubman called in his cutting-edge marketing firm, Yankelovich, Skelly & White, to find out why the A&W campaign was failing. A study had shown that, without question, respondents preferred the taste of A&W's burger over McDonald's.

Except for one small glitch.

"Why," respondents asked, "should we pay the same amount for a third of a pound of meat at A&W as we do for a fourth of a pound at McDonald's?" Since three is *less than* four, reasoned more than half of those questioned, A&W was really ripping them off!

And the problem is not confined to hamburger connoisseurs.

Medical professionals, it turns out, aren't immune to fallacious math either. Doctors and nurses have also been known to err when calculating dosages for medications. The problem is prevalent enough, in fact, to support services that help simplify math for doctors and nurses, including eBroselow.com, whose tagline is "Taking the math out of medicine."

MATH AS A WINDOW INTO THE FUTURE

A nation's future is tied to the educational level of its citizens. I doubt that many would debate that. The national focus on math disparities isn't about the cost of hamburgers at fast-food restaurants, nor is it about sporadic physician head-scratching about dosages. What it *is* about is the level of achievement attained by the students who will someday grow up to be an integral part of determining this democracy.

That is where the concern is intense. And warranted.

On the Program for International Student Assessment Examination (PISA), which ranks math achievement in high school–age students throughout the world, in 2012 the United States ranked . . .

1.

2.

3.

4.

5.

6.

7.

8.

9.

10.

11.

12.

13.

14.

15.

16.

17.

18.

19.

20.

21.

22.

23.

24.

25.

26.

27. UNITED STATES

28.

29.

30.

31>65

We're in a small boat with Russia, Hungary, and the Slovak Republic.

Who's at the very top? Shanghai, Hong Kong, Singapore, and

Taiwan. According to the study, the math performance of fifteen-year-olds in Shanghai was "over two years . . . ahead of those . . . in Massachusetts," even though Massachusetts is one of the "strong-performing U.S. states."

The somewhat comforting suggestion that our low scores were just the result of a high proportion of poor performers pulling down our average doesn't work. The United States also has significantly fewer "top performers in math." For instance, fewer than 9 percent of U.S. students scored "advanced" in math, compared to a whopping 55 percent of students in Shanghai, 40 percent in Singapore, and more than 16 percent in Canada.

The lag in math scores in fifteen-year-old Americans can be traced, in stages, back to eighth grade, then to fourth grade, then to first grade and even kindergarten. Chinese children, on the other hand, have been shown to excel early on in math, in addition, subtraction, counting, and even the ability to position a specific number correctly on a line between 0 and 100. Chinese kindergarteners have been found to have skills similar to U.S. second graders for number estimation.

U.S. secretary of education Arne Duncan's first recommendation, in response to the PISA scores, was that the United States must seriously begin to "invest in early education." His recommendations included raising academic standards across the board, making college affordable, and doing more to recruit and retain top-notch educators. But his top recommendation was for improved education for children between birth and five years of age, the years critical to the academic achievement, including math, of students throughout their lives.

WHY DO SOME FIND IT EASIER THAN OTHERS?

Why do American children lag so severely in math? Why do children in China and other Asian countries excel? How can we improve?

While precise answers have yet to be determined, there are important areas to explore. It has been suggested, for example, that the early grasp of math in Chinese children is because of the Mandarin language. In Mandarin, for example, the number eleven is ten-one, a logical next step from ten. In addition, support for math in Asian countries, by parents and teachers, appears to be notably different.

Early math research, similar to language research before Hart and Risley, was less about finding the reasons for differences in math proficiency and more about exploring the universals of math development in all children. At that time, it was generally accepted that children came to school as "mathematical blank slates," ready to absorb math according to individual innate ability. Jean Piaget, the highly influential developmental psychologist whose "theory of cognitive development" influenced educational pedagogy, actually believed that math should be excluded from early childhood education on the grounds that young children were "preoperational" and not ready for abstract mathematical thinking.

"Average children of four or five years may be able to count . . . up to, perhaps, eight or ten," stated a staunch follower of Piaget, "but the illuminating . . . experiments of Piaget . . .

show that behind this verbal façade these . . . children have . . . not a glimmering . . . idea of numbers."

It was only when researchers began to study young children, then toddlers, then infants and, eventually, newborns that they began to find much more than a "glimmering idea of numbers." Amazingly, they were finding mathematical competence from almost the first day of life.

Contrary to Piaget's theory, it turns out that babies come into this world with an innate, non-verbal "number sense" and the ability to "guesstimate" the relative number of things. Newborns just two days old are, in fact, even capable of doing a kind of numbers matching game. Researchers found that when they played a specific number of syllables to newborns, the babies were able to match it to the correct number of geometric shapes. For example, when the newborns were played "tuuuuu tuuuu tuuuu tuuu," they would look longer at a picture with four squares; when twelve syllables were played, they looked longer at a picture with twelve squares. Even more impressive, the ability of infants to link the number of sounds to the number of objects at six months of age often predicted their eventual math ability.

THE APPROXIMATE NUMBER SYSTEM

This "approximate number system" is the first stage in our ability to work with numbers. It refers to the ability to estimate numbers and then perform basic mathematical procedures related to those estimates.

As adults, when we're given a choice of several jars of M&Ms, unless we're on a strict diet, we home in on the one with the most candy. Even if we're on a strict diet, for that matter. When we're at the supermarket and there are ten lines, we quickly assess the length of each, then head to the shortest, adroitly cutting off someone else who had calculated this at the same time. In both cases, we are using our approximate number system. Before we start feeling too smug, it should be noted that this isn't an ability unique to humans; we share this innate sense with rats, pigeons, and monkeys, among others.

Unfortunately, while our innate sense of numbers would seem to put us on the right trajectory for understanding the words associated with those numbers, it doesn't. Which is, it turns out, very important.

THE CARDINAL PRINCIPLE
NOT AS EASY AS ONE, TWO, THREE

Even with the approximate number system in place, it's a long road from the newborn's ability to assess numbers to learning algebra, calculus, and higher math. And that is where, science strongly indicates, the early language environment again becomes vital. Because while the approximate number system gives us the early ability to intuitively estimate numbers without relying on words or symbols, the ability to proceed to higher levels of mathematics is absolutely language dependent.

Cheerios, as many parents know, is not simply a cereal; it's an early method of teaching numbers. "One, two, three, four, five," I'd say to my youngest child, daughter number two, as I pushed Cheerios across her high-chair tray. "One, two, three, four, FIVE!" Then one-year-old Amelie, blithely unconcerned about her acumen in mathematics, would repeat, "One, two, three, four, five." Well, not actually "one, two, three, four, five" but, to a mother, a reasonable approximation. "Very good," I'd say, supportively. And she would smile and I would smile and her brain, rapidly accumulating skills and strengths, would store away the numbers and the words for the numbers on her rapid, edible way toward calculus.

Like Amelie, almost all young children can repeat a string of numbers: "one, two, three, four, five." And when they do, we beam at them, our very own budding Einsteins. But it's a very long journey to understanding that those words refer not simply to individual things but rather to a "set" of individual things.

That is, when a child counts "one, two, three, four, five," pointing to an individual Cheerio for each number, it actually may seem to that child that each number refers to one thing. It's a quantum leap to understanding that the word "five" is actually an abstraction for the five things as a group . . . five Cheerios, five rabbits, five fingers. Grasping that fact, that numbers represent individual things *in a group,* whether there are two or twenty-two, indicates an understanding of what is called the "cardinal principle." When this concept is understood, it is an important indication that the child is on a path toward understanding higher mathematics.

Grasping the cardinal principle is optimally complete at about four years of age. Why is that so important? In addition to his other important work, Greg Duncan, distinguished professor of education at the University of California, Irvine, has demonstrated that the math skills of children at school entry were predictive of math *and* literacy achievement through the third grade and of math to age fifteen. While innate math potential may play a part, disparities in language environments in the first three years appear to play a significant role in determining which children will have the math skills at school entry that will set them on a positive trajectory in math.

WHAT REALLY COUNTS

PARENT MATH TALK

Professor Susan Levine and her colleagues at the University of Chicago, following approximately forty-four fourteen- to thirty-month-old children and their families as part of the Language Development Project, have added immeasurably to our understanding of the importance of the early language environment for overall cognitive development. They painstakingly videotaped every word, gesture, and interaction of parents and children at home during study sessions, and their research has confirmed Hart and Risley's findings of the tremendous importance of a child's early language environment for preparing a child for later school success. But Levine and her team have uncovered even more nuanced and powerful effects of parent talk.

As Susan Levine looked closely at the transcribed videos, she discovered that the expected variation in quantity and quality of words was compounded by an immense variation in math talk.

During five ninety-minute home visits, some children heard about 4 math-oriented words while others heard more than 250. In one week, therefore, some children would hear 28 math words, others 1,799. Extrapolating to one year, this meant the 1,500 math words one child heard could be contrasted to almost 100,000 for others. An enormous difference.

Professor Levine and her colleagues then used a test to determine whether those differences were predictive of a child's math ability. Children just under four years of age were shown two cards, each with a different number of dots. The children were told a number then asked to point to the card with that number of dots on it. The researchers wanted to know, of course, if the children were able to match the *word* for the number to the actual number of dots.

There was no doubt. Children who had been exposed to more math talk were predictably more likely to choose the cards with the corresponding number of dots. Their superior understanding of the essential cardinal principle of mathematics, compared to peers who had heard much less number talk, corroborated the power of parent talk.

SPATIAL ABILITY

A derivative of the word "space," spatial ability, another math-related skill, refers to understanding how things relate to each other physically. For example, the distance of the sun to the earth, the way a puzzle piece fits into another puzzle piece to help form a picture, the difference between the ground floor of the Empire State Building and the 102nd. It also refers to visualizing spatially, deciding the right direction to take, and even, as we have seen in the concept of DNA, Watson and Crick's re-forming Rosalind Franklin's flat image into a three-dimensional model that became the famed double helix.

Aaron Klug's 1982 Nobel Prize acceptance speech, in which he thanks Rosalind Franklin for her two-dimensional images that he and his team were able to configure into three-dimensional constructs of nucleic acid proteins, is an example of how spatial intelligence allowed genius to compound genius.

Spatial ability, which is an important predictor of achievement in science, technology, engineering, and mathematics, also appears grounded in parent talk. In her study, Susan Levine examined how parents' "spatial talk" differed in the use of words indicating sizes and shapes of objects, for example, circle, square, triangle, larger, round, pointed, tall, short, etc., and whether these differences affected a child's understanding of the spatial relationships between objects.

The results were impressive. During the two and a half years of the study, which began when the children were fourteen months old, the discrepancy in the amount and type of spatial talk each child heard was significant, with some hearing as few as 5 spatial words and others more than 525 in the thirteen and a half hours of recorded time. It's not surprising that children who *heard* more spatial words were more likely to *speak* more spatial words, with an incredible range of 4 to about 200.

Two years later, when the children were four and a half years old, the team assessed them again. This time they looked at their spatial *skills,* including how well they could mentally rotate objects, copy block designs, and understand spatial analogies, the "what-ifs" of spatial knowledge.

Again, the results were not surprising. Professor Levine and her team found that children who had *heard* more spatial words and *used* more spatial words performed much better on the spatial tests. The data showed that this was not just because of being "smarter," but that it was entirely related to their experience with spatial words.

Professor Levine's work demonstrated that a concrete, nonverbal ability can develop as a result of words. The question, of course, is how? Does hearing about the physical design of an object and its relationship to other objects increase a child's awareness of spatial design and spatial relationships universally? To me, this is simply another example of the incredible power of the brain to translate words beyond the ideas they communicate into broader and more complex comprehensions and abilities.

An important "however." While feeding a child's brain the right "knowledge nutrition" is an effective first step to helping that child understand a subject, including math, all children who understand spatial relationships at four and a half will not become an Einstein or a Tesla. Just as a great potential pianist may still be playing "Chopsticks" thirty years down the road if his or her response to "It's time to practice," is "Later, Mom," so, too, a child who has excellent spatial skills at four and a half, but who'd rather play football or write short stories than do math, will probably never become a mathematician. The foundation is there, but so must be the interest, the practice, and the practice even more.

GENDER DIFFERENCES
HOW SUBTLE INFLUENCES
CAN HAVE PROFOUND EFFECTS

Early math talk as a catalyst for ultimate math achievement may have traditionally bypassed girls. While the results have not been confirmed by all studies, in one study of middle- to upper-middle-class mothers, daughters under two years of age received half the overall math talk than did sons. In addition, girls in that study received approximately one-third of the essential "cardinal number" talk than that spoken to boys. While not all other studies have shown a gender difference in early math talk exposure, there is likely a more powerful type of talk impacting math achievement in girls: gender stereotyping. This may well be a factor that shifts girls away from fields that might interest them, preventing them from developing an expertise in areas that might benefit from their participation, including the important STEM fields of science, technology, engineering, and math.

Studies indicate that this problem may begin in the first stages of life, when a parental and societal preconception of math ability in girls translates into a lack of encouragement and subtle discouragement. Girls who are told, even subtly, that math is "not their thing," often, in very concrete ways, do not do well in math.

How does that happen? Isn't a natural ability there to call on at will? No. Words, as we know from their impact on self-image, also impact skills. When your self-image is as a math "nonperformer" and you're challenged to learn a math skill, your brain uses up your intellectual energy by arguing with you that you really can't do it, a kind of mental barricade on the road to accomplishment. You may inherently have the ability to learn it, but that ability is eroded by the roiling doubts that wear it away. Even girls who do well in math typically assess themselves as inferior to their male counterparts, self-stereotyping apparent in girls as young as seven. Shown to affect long-term achievement, it's hard not to equate this with the relatively small numbers of women going into math, engineering, and computer science.

Recent research indicates, however, that this may be changing. Studies now show a narrowing of the American gender gap in math achievement and that increasing numbers of girls do as well as boys in math. The number of females working in the STEM fields is also increasing. Importantly, this may well be attributed to a change in preconceptions about gender-determined math ability and a concomitant more positive approach to girls at home and in school related to math learning.

One of the greatest ironies is that the gender stereotypes seem to be a legacy handed down from mother to daughter, a reflection of one generation's insecurity being passed on to the next generation, ad infinitum. Moms consistently overestimate their sons' math abilities while underestimating their daughters', even when confronted with actual math achievement. There is also more of a tendency to get sons involved in math activities, influencing both actual participation and interest. In addition, mothers have been found to more often predict that their sons, in contrast to their daughters, will be successful in math-related careers. Most astonishing, and sad, is that this is done even in the face of actual academic achievement, with girls internalizing it, no matter how well they're actually doing in math. When girls succeed, the philosophy tells us, we and they know intuitively it's because they "worked hard"; when they fail, it's because of a "lack of ability." In contrast, when boys succeed, it's because of their innate ability; when they fail, it's because they "didn't work hard enough."

Professor Sian Beilock, the author of *Choke,* studies the impact of stress and anxiety on learning and performance in all professions. She and Professor Susan Levine, in their research on learning, have discovered another powerful example of women passing on their own math insecurities to girls. Their study looked at the effect of the preconceptions of elementary school teachers on math achievement. In a profession that is 90 percent female, only 10 percent have math backgrounds and, as a group, they tend to demonstrate the highest rate of math anxiety of all college majors.

The study consisted of seventeen female first- and second-grade teachers and their students. At the beginning of the school year the teachers were assessed for math anxiety. Then the fifty-two boys and sixty-five girls in their classrooms were assessed for baseline math levels, levels found to be completely unrelated to their teacher's level of "math anxiety."

By the end of the year, the level of anxiety that had been shown by individual teachers at the beginning of the year was reflected in the girls in their classes. Girls in classes with "math-anxious" teachers were found, by the end of the year, to be more likely to draw a boy when told a story about a student who was good at math and to draw a girl when the story was about a student who was good at reading. Not only had the elementary school teachers transmitted their own math insecurities in the form of gender stereotypes to some of their female students, the girls who had internalized the negative gender stereotypes did significantly worse on the math achievement tests than boys overall.

On the other hand, girls in classes with teachers who were not "math anxious" were more likely not to exhibit gender stereotyping in math and to score as high as boys.

CALCULATING THE DIFFERENCE

My maternal grandmother, Sara Gluck, the daughter of poor immigrants, propelled herself through the University of Pittsburgh in the 1930s as, you guessed it, a math major. The first in her family to go to college, she did it by working two jobs, switching to Education in her final year because, as my grandfather told me, the only possible jobs for a woman were in education and nursing, another story of gender stereotyping.

What was the difference between my grandmother and the rest of the females of both of our generations? Too late to know for sure. But hearing about her personality, a very strong, determined one, might give us a clue. She's actually someone Carol Dweck might have liked to interview.

CAROL DWECK AND THE GROWTH MINDSET MOVEMENT

Professor Carol Dweck, professor of psychology at Stanford, who authored *Mindset: The Psychology of Success,* is the leader of the "growth mindset" movement, a revolution of thought that has profoundly affected the field of education. Instead of instilling a sense of the absolutes in abilities, says Professor Dweck, what we as parents and educators must engender is the sense that effort is the pivotal factor in achievement, that giving up, not a lack of ability, is usually the cause of failure.

We defeat our purpose, Professor Dweck says, by simply praising innate abilities: "You are really good in math." "Math just comes naturally to you." In that way, we transmit the idea that math is a fixed ability, a "gift" you are either born with or not. Conveying that eliminates the critical importance of perseverance, devotion, and hard work. It implies that when you can't do something easily, you just aren't smart enough, no sense in trying.

In "Is Math a Gift? Beliefs That Put Females at Risk," Carol Dweck's chapter in the book *Why Aren't More Women in Science?,* she eloquently reviews research, her own and others', concerning women's role in the sciences. In it she paints a scientifically confirmed picture that gender stereotyping, ingested by women as true, is the pivotal reason for their poor math performance. She points out that by eighth grade there is already a tremendous difference in girl-boy math grades, *but only for girls who view intelligence as gender specific and fixed.* For those who view intelligence as malleable and improvable, gender stereo-typing, and its effects, almost disappear.

For boys, on the other hand, belief or disbelief in stereotypes has little bearing on their achievement, probably because they are not affected by the stereotypes' negative side.

In addressing the "what to do?" part of the question, Dweck and others asked what would happen if they confronted stu-dents about their beliefs, disproving the question of fixed ability, and convincing them that math achievement isn't a gift but rather the result of hard work? Would this make a difference in their ability to achieve mathematically?

This question resulted in an eight-session program adminis-tered to students in junior high school, a time when math grades often decline and the gender gap becomes most apparent. The experimental group was taught that the brain is like a muscle and that intelligence and expertise are accrued over time. The control group just learned general skills; the question of intellec-tual malleability was not broached.

The results were not surprising to those aware of the effects of gender stereotyping. The experimental group, which had learned that intelligence was an ongoing, developing process, ended the year with significantly higher grades compared to the control group. In fact, in the experimental group, gender differ-ences in math grades almost disappeared. On the other hand, in the control group, girls did significantly worse than boys, strong confirmation of Professor Dweck's theory.

There was another interesting result of the study. Teachers were later asked to evaluate their sense of each student's learn-ing motivation. Without knowing which group students had been in, teachers identified students from the experimental group as having "marked changes in their motivation to learn," reinforcing the potential impact of words on the development of specific skills, including math, as well as on the fundamental drive to learn.

DRIVE AND DETERMINATION

We can add, we can punctuate, we can figure out where we are in the universe. Where do we go now? And how much effort are we willing to put into getting there?

As we've said, prohibitive, negative language with a young child is a barrier to brain development and learning. But does that mean that simply saying, "You're great," "You're smart," "You're brilliant," produces a child who will grow up to be great and smart and can do anything? Turns out . . . no.

There are, it turns out, ways of praising a child that are actually counterproductive. Counterintuitive, too. After all, the reason we bombard our children with "You're so smart," "You're so talented," is because we feel that when children think they're smart, they will be smart. Stands to reason. When you feel good about yourself, you can do just about anything you want.

Right?

No, says Professor Dweck. That type of praise was a post–World War II phenomenon in the United States when, in addition to an unparalleled growth in the economy, there was a radical change in child rearing that contrasted significantly with that of previous generations, when children had generally been expected to "fit in" to the family and rarely did parents revolve around the needs of their children. It was spurred, at least partly, by Nathaniel Branden, a psychotherapist, disciple, and lover of Ayn Rand.

In *The Psychology of Self-Esteem,* Branden expounds on the theory that feeling good about oneself is the key to both individual happiness *and* curing social problems. He hit a sensitive spot in the minds of adults who had been brought up in a much less coddled way.

His message also appealed to John Vasconcellos, a member of the California State Assembly, who set up a special governmental taskforce to "Promote Self-Esteem and Personal and Social Responsibility." Its ultimate goal was to "inject self-esteem" as a "social vaccine" in California to help combat crime, improve low school achievement, and eliminate teen pregnancy, drug abuse, and many other ills plaguing society. Parents and the school system were encouraged to praise children for their intelligence, making them "feel smart," which, it was expected, would motivate them to learn.

This was also a time when every child on the baseball team, win or lose, home run or strikeout, got a trophy and parents anguished that criticizing meant a permanently damaged ego.

In my bookcase is the final report of that California task force: "Toward a State of Esteem." It sits neatly wedged between two other tomes, a bit musty with age, serving as a strong reminder that an idea can excite a population, but if experience doesn't substantiate it and science can't validate it, no matter how good it sounds, its best fate is to sit comfortably, unproductively, forever and ever, on a bookshelf. As good as the self-esteem movement sounded, it just didn't work because, as one critical review stated, the theory of self-esteem "was polluted with flawed science."

Then along came Professor Carol Dweck.

"If praise is not handled properly," Carol Dweck has said, "it can become a negative force, a kind of *drug* that, rather than strengthening students, makes them passive and dependent on the opinion of others."

Professor Dweck's research has pointed us down a very different child-rearing path. It is not self-esteem, the eyes turned inward, smiling with self-satisfaction, that we are aiming for. What we want are children who see a task and, no matter how challenging, almost immediately begin to assess how it can be accomplished, no matter how difficult, no matter how long it may take. When you think about it, that's what parents have always aimed for: stable, constructive, motivated adults. What Professor Dweck's science shows us is that achieving this is a matter of reinforcing assiduous determination rather than inborn ability. What we really want our children to feel about themselves is that, when faced with an obstacle, they can find a way to conquer it by simply not giving up.

It's called "grit."

Grit, the new educational rallying cry, is the trait of tenacity that inspires a child to work hard and relentlessly toward a goal. Angela Duckworth, professor of psychology at the University of Pennsylvania, and Paul Tough, author of *How Children Succeed: Grit, Curiosity, and the Hidden Power of Character,* have been at the center of elevating this idea. While it is not a question of either/or, it is apparent that no matter how smart you are, nor how talented, without the factor of determination, that is, grit, those traits have increasing irrelevance.

Although the importance of grit is undeniable, the science of how to adequately instill it in a child, or even how to measure it, is still young. Nonetheless, the process has begun. Professor Duckworth, who is actively investigating ways to nurture grit, has noted that "children who have more of a *growth mind-set* tend to be grittier." While grit and growth mindset are not perfectly correlated, the growth mindset attitude that " 'I can get better if I try harder' [may] help make . . . a [more] tenacious, determined, hard-working person." Professor Duckworth goes on to say, "[children with a growth mindset are] much more likely to persevere when they fail, because they don't believe that failure is a permanent condition."

Here are the most important differences between being "smart" and being gritty:

When people who think of themselves as innately "smart" can't do something, it's because they weren't smart enough to do it . . . or somebody rigged the problem . . . or it wasn't important to do anyway . . . and they give up.

When "gritty" people can't do something . . . it's because it was only the first try . . . of many . . . and they don't give up . . . not without a real fight. Because they believe, with effort, they can do just about anything.

Intelligence, to innately "smart people," is fixed, unchangeable. "Gritty" people, on the other hand, are simply determined to succeed, which is pivotal in helping them succeed.

The similarity to Carol Dweck's "growth" versus "fixed" mindset is clear. While "growth mindset" believes that "intelligence is enhanced by challenges," a "fixed mindset" believes that abilities are absolute and unchangeable. You are smart or you're not. You can do it or you can't. Often the result of growing up with "gift" praise, "You are brilliant!," "You can do anything," a fixed mindset is a deterrent to continuing in the face of obstacles.

In her pivotal study of 1998, Professor Dweck showed that just a single line of praise, praising either the person or the process, can profoundly influence whether a child is motivated to take on a challenge or not.

In Professor Dweck's study, 128 fifth graders were given a puzzle to complete. After finishing, some children were praised for being smart, others for working hard. The children were then given the choice of a second task, one more difficult, but from which they "would learn a lot," or one similar to the first. Sixty-seven percent of the kids called "smart" chose the *easy* task; 92 percent of those praised for working hard chose the more difficult task.

A robust research literature has emerged reaffirming Carol Dweck's pioneering work, confirming the striking differences in the outcomes of "person-based" versus "process-based" praise. Children with a fixed mindset, attributed to person-based praise, have been found to be more likely to give up when things became challenging and, even more important, were more likely to compound failures by continuing to perform poorly after a failure. They were also more likely to lie about achievement in order to appear smarter, since being thought "smart" by others was such an important part of their personas.

PRAISE IN THE FIRST THREE YEARS

My mentors and colleagues at the University of Chicago, Professors Susan Levine and Susan Goldin-Meadow, teamed up with Professor Carol Dweck to study the effects of praise in early childhood. Led by Professor Liz Gunderson as part of a longitudinal language development project at the University of Chicago, they examined the types of praise children one to three years of age were receiving from their parents. Five years later, they followed up to see if the children's mindsets, growth versus fixed, could be correlated with the type of praise they had received.

The relationship was impressive.

The first stage of their study had shown that by the time children were fourteen months old, parents had already established a "praise style," that is, praise for "smartness" versus praise for effort.

Five years later, they found that children who had received a higher proportion of process-based praise, that is, praise for diligence and effort in the first three years, were much more likely to have a growth mindset orientation to life at seven to eight years of age. Even more compelling, they found that growth mindset predicted math and reading achievement from second to fourth grades. These children, evidence indicated, were prone to believe that their successes were the result of hard work and overcoming challenges and that their abilities could improve with effort.

Disturbing gender differences in praise were also seen, however, although not consistently. In those studies in which gender differences were seen, boys were more likely to receive process-based praise, while girls were more likely to be praised for their "innate" abilities, even at fourteen months of age. Girls in those studies were more likely to think of themselves with a fixed mindset and a concomitant feeling of unalterable abilities. Studies to corroborate these findings are ongoing.

While the results of gender differences in praise have to be corroborated, the effect of process-based versus person-based praise seems clear. In both instances, parents actually have the same positive child-rearing intentions, but doing it successfully, science tells us, is more likely with process-based praise.

GRIT VERSUS GRIT

So, when we talk about grit do we mean any old kind of grit? Or is there grit that we like versus grit that we could do without?

I asked Shayne Evans, CEO of the University of Chicago Charter School, "On your way to eliminating the 'belief gap,' the engrained feeling in your students that they just can't achieve, do you try to instill grit?" Was the belief gap, in fact, just another example of the lack of grit?

Not at all, Shayne explained. The kids in his school have lots of grit, he said. And they do a lot of things with that grit; just not always the things that will enhance their lives academically. What the kids in his school needed, explained Shayne, was redirection of their grit. And that, in fact, is what Shayne and his colleagues help them to do.

DO THEY HAVE GRIT?
IMAGINE DOING THIS

Imagine having to get to school every day by hopping on and off multiple buses in crime-ridden neighborhoods. Imagine dealing with a society that looks at you negatively, even without knowing who you are. Imagine looking at the future and seeing a dark, impenetrable curtain hung in front of it with a small doorway that allows almost everyone in except you, or people "like you." Imagine coming from an inadequate, unequal education system and health care system and being given unequal work opportunities; in other words, imagine living an unequal life. Do these kids have grit? A lot of it, says Shayne Evans, how could they survive without it?

Redirection of their students' grit, to Shayne and his colleagues, means inspiring students not only to finish high school but to go on to college and a life in which they accomplish their goals and dreams. This is predicated on the students' establishing a growth mindset, an internalized feeling that they can do it; it's just a matter of setting their minds to it. Ultimately, it's about taking back the power from a world that has hammered, and continues to hammer, into them every reason to not believe in themselves, and using that power for achievement.

Shayne's thoughts are corroborated by scientific research. Studies have shown that instilling a "growth mindset" in minority students is an important factor in combating the threat of self-stereotyping that has been shown to be a serious factor in academic performance. In an intervention geared to testing the effects of negative self-stereotyping on achievement, minority students in the group taught to view intelligence as malleable had a higher grade point average than the control group, reducing the racial achievement gap by 40 percent.

Between 2012 and 2014, the percentage of graduating seniors at the University Charter School about to enter college was 100 percent: the entire graduating class from each year.

THE ESSENTIAL ELEMENT

Intelligence, a growth mindset, and grit are important factors in achievement. But without another key element, they are, as some researchers have described it, nothing more than whirling dervishes, accomplishing nothing.

Let me introduce you to James Heckman, professor of economics at the University of Chicago and 2000 Nobel Prize laureate in Economics. Professor Heckman's research has demonstrated the amount society saves by investing in the early lives of children. For every dollar invested in the first years of a child's life, Professor Heckman calculated, society would get a return of about seven to eight dollars. Without question, a great investment.

Reflecting his life's goal of determining ways to reduce inequality and promote human development, as well as understanding better how to ensure that individuals can reach their full potentials, in 2014 Professor James Heckman founded the Center for the Economics of Human Development at the University of Chicago. The center's projects include evaluating early childhood programs, looking at strategies that encourage parental investment in their children, measuring and fostering non-cognitive skills including honesty and persistence, and identifying the relationship between genetics and the environment. The ultimate objective of the center, including Executive Director Alison Baulos, MBA, MSW, and the center's fifty committed researchers and staff, is learning what elements are more likely to optimize life outcomes such as educational attainment, employment success, health, and parenting.

When I first met with Professor Heckman, his outer office was a hubbub of PhD students from around the world. He was very impressive: tall, a great shock of white hair, and absolutely charming. He ushered me into his office, motioned me to a chair across from his, and we began to talk. Or, rather, I began to talk. His attention was intense, focused, almost like a computer pulling in data and analyzing it at high speed.

When I stopped talking, he leaned back and eloquently told me what I *should* have been asking him. According to James Heckman, an important determinant of a child's academic success is self-regulation and executive function. Without self-regulation and executive function, there is little chance of achievement in children or, for that matter, in any of us. Ensuring that all children achieve those strengths makes investment in early childhood an important priority.

Sometimes called "character" or "soft" skills, self-regulation and executive function both refer to the ability to monitor and control one's own behavior. Walter Mischel tested these skills with marshmallows.

In the late 1960s, Professor Mischel, a psychologist at Stanford, conducted an experiment testing a child's ability to wait for a bigger prize or take a smaller one now. Sometimes the reward was one marshmallow versus two marshmallows. Decades later he published *The Marshmallow Test,* a book that included the results of his study. These results indicated that children who were able to wait for a larger prize were more likely, years later, to do better academically.

Self-Regulation And Executive Function
The Prefrontal Cortex

Being able to delay taking a marshmallow is really a metaphor for far more important behaviors, including holding back explosive and inappropriate outbursts, controlling responses to temptations, and restraining violent reactions such as screaming in anger or hitting others. Another name is "inhibitory control," that is, inhibiting our "natural" responses when they are negative or intensify a problem.

Separate from our intelligence, executive function and self-regulation keep us steady while we attempt to solve a problem rather than provoking us to spontaneously react in ways that would exacerbate it.

These skills, so essential to a productive, stable adulthood, are not gifts that come from birth. Highly influenced by the environment of our early childhoods, they are acquired and refined over a prolonged period, from infancy through early adulthood, tied absolutely to the part of the brain known as the prefrontal cortex. And therein lies the great importance of the home environment. Because the prefrontal cortex doesn't just develop in a positive way on its own, becoming the perfect epicenter for self-regulation and executive function. If that were true, life would be much easier. In truth, from the time we're born, this part of the brain is highly susceptible and reactive to anxiety and threats. An emotional, toxically stressed early environment that includes, but is not limited to, negative and volatile parent talk, adversely affects the development of the prefrontal cortex, stunting self-regulation and executive function, eventually compromising the ability of the child, and eventually the adult, to deal with the stresses of life.

A child entering kindergarten with poorly developed self-regulation and executive function, for example, will have difficulty learning. If a child's mind can't, in a sense, quiet itself, or concentrate on the information being presented, that information will not be absorbed by the child. It's that simple. The result is not only curtailed learning at that moment, but a poor prognosis for future learning, regardless of the child's potential IQ.

And the disturbance is not limited to the child; there is also an effect on the class as a whole since that child's behavior can disrupt everyone else's activities. The result, in order to diminish the effect, is often to shunt the child aside as "dumb" or "bad," a label that is often both indelible and prophetic.

While all kids are susceptible, statistics have shown that children born into poverty, particularly boys, are especially at risk. Why? Many potential reasons. Poverty itself, with its lack of hope, its draining complexities, is stressful. The birth of a child, a stressor in even the best of situations, can exacerbate this. In addition, environments in which the poor often live are also filled with high stress, including the potential for violence right outside the door.

The effect on the child is not surprising. While stress exists in all lives and, at occasional low levels, may have positive effects, children exposed to chronic, toxic stress are more likely to enter kindergarten already burdened with self-regulation and executive function issues, issues that often continue throughout a child's entire academic life and workforce trajectory.

Understanding why this occurs is important.

When a home is chronically stressed, when verbal communication is harsh, accusatory, and threatening, the "haven" for a child's brain is "hypervigilance," that is, being on constant guard against imminent attack. Sometimes referred to as "fight or flight," this ready-to-defend system is simply the brain's attempt at being self-protective. The problem is, it's so protective that ultimately the brain loses the ability to distinguish between imminent threat and no threat at all. All of its energy is spent in being on guard against the unknown, severely impacting its development.

This loss of brain growth, the result of a total preoccupation with self-defense, results in a severely diminished capacity for abstract learning, including things as basic as the ABC's and "one and one make two." It's a loss compounded year by year as these children, even as teenagers and adults, continue to fall more and more behind their peers. Their actual potential? We'll never know.

A Child's Key To Self-Regulation

Parent talk can influence a child's self-regulation and executive function, but more important is the child's doing it without outside prompting. For example, when we say to a child, "Use your words," we're really telling that child to stop a behavior and to self-regulate. Actual self-regulation, however, is dependent on the child's actually telling it to him- or herself. This is critical, because while there will always be a need for behavior control, it's only when it occurs naturally, without the need for outside directives, that the brain is kept clear for intellectual growth.

Does telling a child to "use your words" always transform that child into someone who does use words instead of less positive responses?

Sometimes, yes. Sometimes, not so much.

Lev Vygotsky, a Russian psychologist who died in 1934 at the age of thirty-seven, was at the forefront of understanding the development of children's self-regulation. Vygotsky proposed that a child's self-regulation develops by way of caregivers who transmit cultural norms in day-to-day interaction with the child, ultimately providing the child with the brain-dependent process of self-regulation. According to Vygotsky, children go from being "slaves to the environment," that is, at the will of their caretakers, to becoming "masters of their own behavior" with the tools given them by their caretakers. These "tools," according to Vygotsky, are both verbal and non-verbal, although Vygotsky focused on language as the primary one for children learning self-regulation.

Current science supports Vygotsky's hypothesis that language skills play a significant role in a child's self-regulation. Children with delayed language, whether due to hearing loss, lack of adequate language input, or delayed speech for other reasons, have a higher incidence of problems related to self-regulation. The reverse is also true. Interventions that focus on vocabulary development have been shown to increase both a child's language *and* his or her social skills. An intervention to improve preschoolers' language skills found that participation predicted improved social skills into early adolescence. Amazingly, the greatest positive effect was in boys, who struggle more with self-regulation, and in children from high-risk families.

Talking To Yourself

Children between two and seven years of age often chat away, no one else in sight. And it's a good thing. It turns out that a key mental tool in children's self-regulation is talking to themselves. Private speech among preschool children, also known as "self-talk," is actually predictive of greater social skills and fewer behavioral problems. Teachers rated these children higher for self-regulation.

The reverse is also true. Children from disadvantaged backgrounds, such as those who participated in a study in Appalachia, have been shown to have relatively poor, less developed forms of private speech, and concomitant negative outcomes for self-control and social skills.

Professors Clancy Blair and Cybele Raver at New York University, in a step toward testing the effectiveness of a program to improve self-regulation and executive function in children, conducted a rigorous, controlled study of the Tools of the Mind program. Their landmark research, which encompassed twenty-nine schools and 759 kindergarten children, illustrated positive effects on executive function, reasoning ability, control of attention, and even levels of salivary cortisol, the hormonal indicator of stress. Their results also demonstrated improvements in reading, vocabulary, and mathematics that increased into the first grade.

Some of the effects were specific to needs evident in high-poverty schools, suggesting that focus during early elementary school on executive function and associated aspects of self-regulation holds promise for helping to close the achievement gap. Even Professor Blair was amazed by the study results. "We were finding that the classrooms in high-poverty school districts that had implemented Tools of the Mind became identical in a wide range of essential specifics to those in high-income school districts."

How Parent Talk Influences Self-Regulation

Parent-caregiver language plays a central role in a child's ultimate ability to regulate behavior and emotional responses. When a child is in a language-rich environment, the increased language skills of the child result in an increased ability to self-regulate. The reverse is also true. The decreased language skills of children in homes with less parent talk translate into a diminished ability to self-regulate.

Emerging research tells us that this occurs even when a baby is too young to understand language. Just *hearing* the natural sequence of sounds, it turns out, puts a baby on the path to self-regulation and executive function. This is because, during the process of learning a language, the brain, hearing a series of sounds, actually begins to develop a framework for processing things sequentially, which, in turn, is a precursor for planning and executing reactions, an important aspect of executive function and self-regulation.

The research was done at Indiana University, where Professors Christopher Conway, Bill Kronenberger, David Pisoni, and their colleagues studied children born deaf who had received cochlear implants. Their conclusion: While *hearing* language impacts far more than language skills in a child, *not* hearing sound affects executive function and self-regulation at an even more fundamental, more profound, level.

Optimal caretaker language, in the very early years of a child's life, is geared at helping a child toward independence. Every bit of praise, every effort toward supporting or correcting, is a conscious, or sometimes subconscious, strategy for getting the child to be independently "good" and independently productive. As in all aspects of child rearing, success is often predicated on sensitive and responsive caregivers who can help children practice age-appropriate behavioral skills and problem solving, perhaps just slightly above what they can do on their own.

Lev Vygotsky called encouraging children to behave just above the edge of their ability "the zone of proximal development." A way of easing children to a higher level of behavior, it is the difference between saying to a young child, "Put the toys away now," versus saying, "What should we do with the toys now that we're finished playing with them?"

The first is the easier, a demand by a "superior" that must be fulfilled, no questions asked. The second, however, supports a child's emerging autonomy; its effect on a child's self-regulation and executive function, supported by science, is enormous. One-year-old babies whose mothers calmly suggested, rather than dictated, rules for behavior, were found at age three to have notably stronger executive function and self-regulation.

The research, done by Professors Grazyna Kochanska, Nazan Aksan, and others, also demonstrates that self-regulation in children is enhanced when parents support their children's control of their own behavior, when they explain reasons for rules, and when they provide non-emotional reasons for discipline. Those children were more likely to think through problems without displaying immediate reactive behavior. The thought is that children internalize parents' management styles as their own "private speech," and these become the basis of their own behavior.

The other side of the discipline coin is the negative impact of parents who are more controlling. Parents who use pressure and authority to restrain their children's behavior may elicit short-term obedience but, in the long run, they also set the stage for poor self-regulation and executive function, producing adults who may have serious problems with self-control.

The Nuances Of Parent Language

The two major categories of regulation are:

> **Directives:** commands that restrict a child's input, including reprimands and demands
> **Suggestions and Prompts:** eliciting a child's input, opinion, or choice

At the very instant a parent yells a directive, the thought is not necessarily how the words, or the tone of voice, are going to impact the adult the child will become. For example, when a mother screams, "Get down from the roof NOW!," the immediate concern is for the child to reach adulthood, period, the question of the child's self-regulation postponed for the moment.

Nonetheless, science does show us that of the two categories of parental speech, "suggestions and prompts" are supportive of long-term self-regulation skills, while a preponderance of directives impairs it.

The science is not quite as clear in the moderate use of directives. It does not, in fact, categorize directives as an absolute negative. Early on it seems the direct and clear nature of directives may facilitate a child's ability to learn rules and develop appropriate behaviors that are relevant to emerging executive function and self-regulation skills.

As in all aspects of human development, generalities are secondary to complex interactions between the individual child and the environment. None of us begins life as a completely empty canvas waiting for our world to tell us who we are, what we can do, who we can be. This is especially true in the development of self-regulation and executive function. It is not just that our genetics and our "at birth" temperament play a role in their development, which they do; they also help determine how we respond to the environment we're presented with. For example, children who appear from birth more reactive or "temperamental" are thought to be hypersusceptible to their environments. This means that in a controlling or hostile environment they grow even more reactive, exhibiting poorer self-regulation. On a positive note, however, in highly supportive environments, some research finds that these children thrive.

Even without the absolutes of science, there should be no question that the optimum environment for a child is warm, nurturing, and responsive. And that for all children, a stressful, toxic environment is negative, inhibiting the development of executive function and self-regulation, affecting both the child and the adult that child will become.

Give Me Your Huddled Masses
The Bilingual Advantage

As a third-generation American, I was privy to a lot of tales, always secondhand, of what it was like to "come over." My great-grandfather arrived in the United States at twelve to spend ten hours a day rolling cigars in Pittsburgh. As my mother says, "They crossed a wide, bumpy ocean in the lowest, rattiest part of steerage to exchange poverty for poverty."

My great-grandfather, his father and, in fact, eventually the entire family came here without skills, without money and, above all, without knowing ten words in English. But by the time my mother lived with them as a child, forty years later, they spoke only one language: English. She said that except for a few phrases in their own native tongue, mostly snippets of clever sarcasm, there was no attempt to speak anything else to the "kids." In fact, it was thought a very bad idea if the "kids" spoke or heard anything but English.

They were wrong . . . but I guess it's too late to tell them now!

In the recent science exploring the pros of speaking more than one language, some studies have found that children who speak a second language have enhanced self-regulation and executive function. This science contradicts the "traditional wisdom" of studies done before the 1960s, indicating that bilingualism negatively impacted intellectual development and IQ. Interestingly, and leading to a question of cultural prejudice, this overall view did not seem to include French, which has always been in vogue. *Bien sûr!*

The flaws in those studies were revealed in 1962 by Professors Elizabeth Peal and Wallace Lambert. Using standardized measures and accurate sampling, they found that being bilingual indicated both a verbal and a non-verbal advantage over those who spoke only one language. Peal and Lambert spearheaded an explosion of scientific literature, including research demonstrating the positive impact of being bilingual on executive function. Initially this was attributed to a baby's having to actively inhibit one language while discerning meanings in another, helping the brain to ignore distractions and focus. But it appears to be more complex and nuanced than that. In fact, those who are bilingual always have both languages at hand, their brains constantly monitoring which one to use.

"We might expect that bilingual speech would be riddled with errors, that sometimes you slip and the wrong language comes out," says Professor Ellen Bialystok, a leading researcher in the field. "But that doesn't happen." Researchers believe that the bilingual brain is always ready to be active in both languages, probably to make sure the response to a Peruvian grandmother isn't mixed up with one to the cute kid in its algebra class. As with self-regulation, the bilingual brain is continuously monitoring for the appropriate response to input. Works with languages; works with life.

Unfortunately, the credo of a hundred years ago persists. Although we are approaching an important milestone in American history, where the majority of our population will soon be of Hispanic heritage, there are still immigrant parents who want their children to speak only English. Just as I suspect my great-grandfather believed, they cling to the idea that English, the language of this country, is the only one necessary for their "kids."

Iara Fuenmayor Rivas, TMW's bilingual curriculum developer, discovered this when she spoke to the immigrant parents who were to become part of our Thirty Million Words–Español research. It surprised even her. The parents absolutely understood that they were their children's first and most important teachers and that their words were instrumental in helping their babies' brains grow. They comprehended and enthusiastically embraced the science.

Except for one part.

As a group, they refused to accept bilingualism as a positive aspect of their children's development, often rejecting the idea of speaking to their children in their native tongues. Even explaining the potential effects on self-regulation and executive function didn't change their minds. To them, their primary goal was that their children become "real Americans" and, to achieve that, English should be their only language.

I can see my great-grandfather "Pup Lewinter" nodding in agreement.

Even though he's wrong.

Erika Hoff, professor of psychology at Florida Atlantic University, is an expert on the impact of bilingualism on children's language development. The babies she's been following from infancy, born into bilingual homes, are just turning five.

This is what her research has shown: that no matter the education level of the parents, nor how proficient they are in English acquired as an adult, it is always better when parents speak to their children in their native tongues. The reason is logical. Since the new language, in this case English, was learned by parents as adults, their proficiency will never match that of their native tongue in vocabulary, syntax, nuance, or overall quality. This is because when people express themselves in the language that's been part of their entire lives, they express more than just the concrete meanings of words; they express the more profound meanings, both emotional and, to the non-native speaker, somewhat veiled. Some studies have even found that learning language from a parent using a non-native tongue negatively impacts overall cognitive development at twenty-four months.

The best scenario is that children of non-English-speaking parents learn the language of their parents from their parents. There should also be a language relationship with native English speakers. While it is true that being bilingual, for a young child, may mean that early vocabulary size in both languages will be somewhat smaller, this is offset by the fact that the child is learning two languages, each of which, as the child gets older, can be enhanced. One of this strategy's greatest strengths is that ultimately these children can have what very few Americans have traditionally had: two languages. And that, to me, is a great strength.

Up to now, we've been talking about the importance of parent talk for intelligence, stability, persistence, self-regulation, and bilingualism. Parent talk influences other attributes that, if everyone in the world had them, would make this planet really, really wonderful.

EMPATHY AND MORALITY
THE SCIENTIFIC APPROACH TO GOODNESS

An important reason for this journey into the power of parent talk is to help design ways to ensure that all children can reach their potentials in life. And those potentials, as TMW's parents constantly remind me, go way beyond academic achievement or professional success. We want our children to be good, not in the "obedient" sense, but in the sense of understanding others with empathy and generosity.

And, being good, it turns out, is a pragmatic decision, as well.

Adam Grant's book *Give and Take: Why Helping Others Drives Our Success* shows that people who are kind and give without demanding reciprocation don't just reap the benefit of saintliness, they often do better in business, too. Grant, a professor at the Wharton School of the University of Pennsylvania, demonstrated that, essentially, "good guys can finish first." This is important not because goodness needs a pragmatic reason, but because it confirms that there are long-term positive effects to goodness.

In his article "Raising a Moral Child," Grant explores the science substantiating parent talk, including praise, as an important influence on a child's generosity and moral behavior. After reading about the positive effects of "process-based praise" on achievement, the answer to, "What kind of praise makes children kind?" might seem to be, "I liked the way you helped your friend with the game." Well, in this case, the carefully collected evidence says no. While helping a child develop persistence in solving problems is achieved by praising behavior, helping a child develop a sense of empathy and kindness is best achieved by praising the individual.

In a study of children who were praised either as individuals or for their behavior, children who had been praised as individuals were, when presented with the opportunity for generosity several weeks later, more likely to be generous.

Another study corroborated this. Children from three to six years of age who were asked to be "helpers" were more likely to help researchers clean up a mess than children who were asked simply to "help." In fact, the children who heard only "would you help?" were no more likely to stop playing and help than children who had heard nothing.

Who knew that a subtle word difference, a verb versus a noun, could change the response of a child in helping with a chore? And it's not just with children. Another study showed that adults were significantly less likely to cheat when they were asked "not to be cheaters" in contrast to being asked "not to cheat." In fact, the group that had been asked not to be "cheaters" were found not to cheat at all.

Why is this true? Probably because most people do want to be "good," and nouns are like mirrors, showing us who we are. Adam Grant describes it this way: "When our actions become a reflection of our character, we lean more heavily toward the moral and generous choices. Over time it can become part of us."

"YOU ARE SO BAD"
VERSUS
"THAT WAS A VERY BAD THING TO DO"

Of course, parent talk is not just about praising good behavior to encourage it. It also responds to behavior that is unacceptable. Guilt and shame are the two ends of the emotional spectrum in response to our having done something wrong. Shame permeates us to the core, describing who we are to ourselves. Guilt, on the other hand, is a pinpoint feeling about a specific action that is contrary to our sense of self: the difference between *being* bad and having *done* a bad thing.

The language a parent uses in response to an unacceptable behavior is pivotal in gearing the child's sense of self in one direction or the other. If we want to help a child on a path of positive actions, criticizing specific behaviors, because they are so different from what one expected, goes a long way toward producing someone who understands that he or she is "good" and just made a reparable mistake rather than someone who now sees him- or herself as "bad."

But in the end, as Professor Grant beautifully points out, there is something even more powerful than the *words* of a parent in rearing a kind, ethical, moral child.

Having a kind, ethical, and moral parent goes a long, long way.

CHAPTER 5

THE THREE Ts

SETTING THE STAGE FOR OPTIMUM
BRAIN DEVELOPMENT

*A person who never made a mistake never tried
anything new.*

—Albert Einstein

PART I
SETTING THE STAGE FOR
OPTIMUM BRAIN DEVELOPMENT

One of my earliest memories, when I first arrived at the University of Chicago in the summer of 2002, is of a skinny graduate student walking toward me, his T-shirt emblazoned with the words "That's all well and good in practice . . . but how does it work in theory?"

It made me laugh to see the U of C reputation, a profoundly theoretical institution, where "practical" was practically a dirty word, showing its humorous side, as well. At least *I* thought it was funny. I was still very grounded in the world of surgery, and this part of the campus, away from the focused lights and controlled rhythm of the operating theater, was a different planet, different culture, different everything. I had as yet to cross the wide green expanse of the Quad into the world of social sciences.

The "practice versus theory" T-shirt is funny only because it has a whiff of truth about it. Too often the theoretical and practical revolve in different orbits, with only an occasional awkward meeting, like a date that seemed right on paper, but it turns out the couple speaks two entirely different languages. Intervention? "Don't understand the word," says the basic scientist. Scientific validity? "Huh?" asks the hands-on interventionist. I can hear both sides of this struggle arguing with me as they read this. But they're arguing in separate rooms in separate buildings in separate academic worlds. Different universes, different languages.

But scientific truths without effective translation into practice won't help our children. And programs designed without a rigorous scientific foundation won't work, either.

This chapter describes the fundamental approach of the Thirty Million Words initiative for translating the science into action. Its goal is the optimum development of children's brains. Its core tenet is based on the malleability of children's intelligence and the power of parent and caregiver language as the critical factors for a child's cognitive development. The TMW initiative includes curricula to be used with parents during home visiting, at well-baby pediatrician visits, and even in maternity wards. Science guides both the development and the testing of all its programs.

While I didn't start off knowing how to build a research program, nor even how to develop behavioral interventions, I knew what my goal was: to understand why some of my patients had more difficulty learning than others, then to design solutions to improve their outcomes. I knew it would take hard work. I knew it would take a team. What wasn't clear at first were the complexities of developing effective behavioral programming. I had a lot to learn.

Scientific papers can be wonderful to read, fun, in fact. Because, by the time we read them, the work is mostly done: the preliminary identification of the problem, insight into reasons for the problem and even, often, the answer to what can be done about the problem. In the medical and technological sciences, professionals and entrepreneurs are waiting at the door to translate that science into action.

This is not true, however, in the world of the social sciences. Even though the work done by social scientists is extraordinary, with rigorously tested, excellent results, those results do not translate as easily into action as the work done in the medical or technological sciences. The reasons for this are complex. Support for ameliorating societal problems takes, rather than makes, money. Initially, that is. Society's answer to "what to do?" is often conjectural, in constant political debate, the proof of credible science no match for "gut feelings." Finally, it's also true that social problems often involve long-standing societal complexity, making innovative, although scientifically supported, solutions often more difficult to put into effect.

That's one of the first things I learned as I became involved in the world of the social sciences: that as much as I had to learn about the intricacies of the science, as hard as I had to work taking its pertinent results and adapting them for the world I wanted to help, it turns out that was the easy part.

It Takes A Really Big Village

The question was never, "Should we?" but *How* do we?"

The TMW initiative is a team of diligent, humane, creative people who work beautifully together. We also work closely with incredible collaborators at the University of Chicago and throughout the country.

Kristin Leffel, director of policy and community partnerships, joined TMW right after receiving her bachelor's in Social Policy from Northwestern University. When we were starting out, on new and unproven ground, Kristin was TMW's everything, including curriculum developer, home visitor, data coordinator, and even graphic designer. In addition to a keen intelligence and creative mind, she is imbued with an extraordinary sense of humanity.

I have to mention that I wasn't Kristin's first choice. Her original letter had actually been to my husband, Don Liu, a pediatric surgeon, asking about work in health disparities research. She had an interest, she told him, in public health and wanted to "make a difference." He passed her letter on to me and the rest is professional history.

Lucky me. Lucky TMW.

My next stroke of good luck was when my brother Michael fell in love with Beth (now) Suskind and induced her to fall in love with him. An established, highly regarded TV producer, she joined TMW as my co-director after much begging on my part. Her astute sense of project design for achieving educational goals has resulted in TMW's honed and accessible curriculum, one that is both easily understood and successfully mastered by our parents.

Thank you, Michael.

Kristen and Beth exemplify the TMW team and its shared vision of trying to solve a critical human problem: what happens to our nation's children.

TMW's Parents
Creative And Collaborative

The TMW curriculum is parent developed, parent tested, and parent directed. Our first research group was made up of mothers and grandmothers who worked in the University of Chicago hospital cafeteria. Generously giving up break times, they reviewed details of the project, then told us what they thought.

Other mothers and fathers were recruited in hospital waiting rooms, in grocery stores, and even at bus stops. Over time, hundreds and hundreds of parents reviewed pages and pages of ever-changing module examples, then gave us their thoughts regarding quality, clarity, and relevance. These comments, critiques, and suggestions from extraordinarily involved parents were invaluable to the curriculum's development. Many a carefully crafted idea, deemed brilliant by our team, was thrown out or completely overhauled because of the astutely critical observations of parents. This process has helped ensure that TMW is not only rooted in science but clear and accessible to those who use it.

Before we discuss the basics of our program, it's important to stress that TMW is founded in rigorous science and, as science guides us, it will evolve. TMW will never be based on what we "believe" to be true or hope will be true. Rather it's based on what we meticulously learn to be true. We are committed at TMW to supporting our theories with well-designed research so that what we do and how we do it can be statistically validated. In the same way, when what we theorize is not substantiated by the numbers, it's either revised or gone.

The only thing we are absolutely committed to is helping parents understand the scientifically confirmed power of their words in building their children's brains. And, in turn, designing programs to help parents use that power successfully. That is the foundation of Thirty Million Words.

Babies Aren't *Born* Smart; They're *Made* Smart By Parents Talking With Them

TMW's foundation is the scientifically demonstrated truth that "babies aren't born smart; they're *made* smart"; that is, the malleability of intelligence.

We are all born with potentials in many spheres, but reaching those potentials is not without effort. Just as seeds have the potential of becoming roses or petunias or hydrangeas, the ultimate beauty and strength of each flower is dependent on the nurturing it gets. Try growing those seeds in a dark basement with little water and you'll understand.

It's the same, it turns out, with the brain. The science showing the development of the brain, including its dependence on its environment for optimum growth, is discussed throughout the book. TMW programs are a result of that science. Our beautifully created animations and videos are based on it. In addition to teaching the essentials of an optimum early language environment, the animations help parents understand that a child's intelligence is not set at birth but is very dependent, for optimum development, on the language environment they provide.

One of our animations demonstrating how "words grow your baby's brain" shows words flowing into the ear through to the brain, where, in a very cute way, they begin to stimulate the brain's neurons. Is that the way it actually occurs? A reasonable animated facsimile. But out of it came one of my favorite TMW stories when, at the beginning of one session, the mom greeted her home visitor by saying, "I think I built a lot of brain connections for my baby this week!" She said it with a smile, knowing it was humorous but also that it was true!

Creating A Rich Early Language Environment

We know the critical importance of a rich early language environment for the brain development of a baby and young child. The important question to us at TMW was how to help parents achieve that environment in order to optimize the benefits for their child. The result was the core strategy of TMW, the Three Ts: Tune In, Talk More, and Take Turns. With the goal of creating an environment for optimal brain development in a baby and young child, the Three Ts translate the complex science of language exposure on brain development into an accessible, easily used program that enhances parent-child interaction while melding well with everyday life.

It's important to stress that establishing a positive early language environment for a child does not simply relate to providing vocabulary; rather, it's reflective of a warm, nurturing relationship. This is not to discount parents who, even if they don't talk significantly, do show affection. But language is definitely a way someone shows us that there is interest in what we are communicating, that here is someone who wants to connect with us in an authentic and positive way.

Creating a rich language environment also does not mean carving out dedicated blocks of time in an already busy life. The Three Ts are designed to become a natural part of everyday activities, no matter how mundane. By adding words, a parent or caregiver transforms making the bed or peeling an apple or sweeping the floor into a brain-building experience. Ultimately, these words will be an important part of enhancing the parent-child relationship as well as a child's brain.

THE THREE Ts	THE THREE Cs
Tune In	Conéctese
Talk More	Converse Más
Take Turns	Comparta Turnos

Whether a parent is talking about the smell of a diaper, or the color of a flower, or the shape of a triangle, the Three Ts are designed to provide the foundation, from the first day of life, for the rich early language environment essential for building a child's brain.

The First T: Tune In

Of the Three Ts, Tune In is the most nuanced. It involves a parent's making a conscious effort to notice what a baby or child is focused on, then, when it's appropriate, talking *with* the child about it. In other words, focusing as the child is focused. Even if the child is too young to understand the words being spoken, even if the focus is constantly changing, Tuning In refers to a parent's following and responding to a child's lead. It represents the first step in harnessing the power of parent talk to build a child's brain. If a parent is not Tuned In, the other Ts will not work.

An example.

Mom or Dad, loving and well-intentioned, sits down on the floor, a favorite children's book in hand, maybe Jolly Roger Bradfield's *Giants Come in Different Sizes,* one of my favorites. Mom or Dad pats the area next to the child and smiles. This is the signal for the child to cozy up and listen. But the child doesn't respond, continuing to build a tower with some old blocks strewn on the floor. Mom or Dad pats the carpet again. "Come on. Sit here. This is a really good book. Daddy (or Mommy) is going to read it to you."

Good, right? Loving mother. Loving father. Great story. What more could any child want?

Well, maybe a mom and dad who are interested in what their child is doing, then joining in, as if the child had patted the carpet and said, "Come on, Mom. Come on, Dad. Sit here. Stacking these blocks is really fun."

In other words, Tuning In.

In the scenario designed by TMW, that's exactly what happens. Parents learn to be aware of what their child is doing, then to become part of it, enhancing the relationship, helping to improve the skills being used in play and, through the ensuing verbal interaction, helping develop their child's brain.

Let me stress why that is important. When a parent plays with a child in the child's area of focus, even if that interest lasts for five minutes and then shifts to something else, the child's brain development is enhanced. That's because the brain does not have to use energy to switch to another arena, specifically one of less current interest. A mom or dad can ask, "How would you like me to read you a book?" That's very positive. What is important, however, is Tuning In to the child's answer, even if it's not verbal, and even if it's not what the parent wants to hear. In its very valuable essence, that is what's meant by Tuning In.

This is better understood when an essential difference between adults and children is recognized. As adults when we're asked to switch directions to a different assignment, we automatically shift focus from what we're doing, even if it's what we'd *like to be doing,* to the task at hand. That's what makes us responsible adults. Children, however, whose executive function is still underdeveloped, stay focused only when they find an activity interesting. If there is no interest, then words, even the words of a really good story, float into the air, having little or no effect on that child's brain development. The effect is on retention of vocabulary, as well. Studies have shown us that when a child has to participate in an activity in which he or she has little or no interest, the child is less likely to learn the words being used.

Tuning In is also enhanced by a parent's being on the same physical level as the child, including joining the child on the floor during playtime, holding a child on a lap while reading to him or her, sitting together during mealtimes, or picking a child up to watch the world from a parent's vantage point.

Conversely, Tuning In is deterred by digital distractions. Computers, tablets, and smart phones are addictive and attention absorbing. Only when the child is a parent's primary focus will the necessary attention for optimum brain building occur.

When the environment is optimal, on the other hand, when a parent follows the child's focused attention and relates to it, communicating in rich, caring language, that is, Tunes In, the parent does more than help a child learn words. A child who receives consistent Tuning In is more inclined to stay engaged longer, to initiate communication and, ultimately, to learn more easily.

Child-Directed Speech

Optimally, Tuning In is a two-way street. Just as babies use sounds to get attention, parents do it, too, changing their tone and pitch to entice and appeal. As we've discussed, "child-directed speech," also called baby talk or parentese, helps a baby's brain learn the language. In a recent study of children from eleven to fourteen months old, those who had heard more child-directed speech knew, at two years of age, twice as many words as those who had been exposed to more adult-directed speech.

But child-directed speech serves another important function in the parent-child relationship. Used by parents with young children throughout the world, in an array of structurally diverse languages, including indigenous languages of Europe, Asia, Africa, the Middle East, and Australia, its melodic pitch, positive tone, simplified vocabulary, and singsong rhythm a few octaves higher than usual entices a child into shared attention. Parents who take pride in never speaking baby talk, speaking to their infants only in the way they'd speak to an adult, are missing an important point: that this kind of talk does not "dumb down" content. By appealing to a baby's ears, it helps to draw attention to what is being said, and to who is saying it, encouraging the child to be attentive, to be engaged, and to interact. In other words, to Tune In.

A key feature of child-directed speech is repetition. To understand the relationship of repetition in encouraging a child to Tune In, researchers at Johns Hopkins University studied sixteen nine-month-olds for ten days during a two-week home-visiting program. During each visit, the babies heard the same three stories, each containing words not normally heard in a baby's everyday experience. A control group heard no stories at all.

After a two-week hiatus, the babies were brought to Johns Hopkins, where they listened to recordings of two different lists of words. The first list contained words taken directly from the three stories; the second recording was of similar, but different words.

Babies who had heard the three stories during home visits listened *longer* to the list of words from the stories. Babies from the control group, who had not heard the stories, showed no preference for either list. The conclusion? Babies "learn" words they hear more frequently and will listen longer to sounds they've heard before, that is, Tune In.

The key purpose of Tune In is parental responsiveness. A child's future well-being, including cognitive development, social-emotional development, self-regulation, physical health, and countless other outcomes, has been linked to the responsiveness of the mother, particularly in the first five years of life. Science tells us clearly that sympathetic, appropriate responses to a child are essential to behavioral and brain development.

While being a parent has long been deemed an intuitive, anyone-can-do-it process, in fact, it isn't. Many exhausted mothers and fathers would agree. Parental responsiveness, the essence of Tuning In, boils down to a three-step process:

1. Observation
2. Interpretation
3. Action

The clues babies and young children use to communicate their needs are both verbal and non-verbal. Ever hear a baby cry? Or, for that matter, a two-year-old? There are few things as attention getting, or heart wrenching.

Interpretation isn't always easy. It is, however, an important precursor to the third step: action, or *what to do?* Is the child tired? Hungry? Bored? Wet? Interpretation, as all parents know, is a honed skill and one that is rarely absolutely accurate. This difficulty in absolute accuracy mandates an overriding element, a correction provision, that should always be present.

No matter what the reason for a baby's or child's behavior, even when the reason for a behavior is not known, no matter what action is deemed appropriate, the key element is warmth. A caregiver's loving, positive responsiveness to a child is an essential factor in that child's development as a human being. No matter what country, what culture, no matter the temperament of the child, handling things in a loving, responsive way predicts the ultimate stability of the child. Science also tells us that.

The Stress Of Parenting

When an infant cries, it might be for any number of reasons, but there is one constant underlying all infant crying: he or she is feeling stress.

And so are the parents.

The pivotal question: What to do?

The pivotal answer: Respond. That's it. Respond.

No matter why a baby is crying in this new and unfamiliar world, which is, let's face it, the scary unknown, the first thing that a baby should understand is that he or she is safe. "Don't worry, little one, Daddy's here. Mommy's here." It's the first lesson in life, a very important one, with long-term effects. *It's not always going to be easy,* you're saying, *but when the times are tough, someone will be there to catch you.*

While some stress may be considered "normal" or even good, constant stress in a child has been shown to have long-term negative ramifications.

What Happens When No One Is There
The Attachment Theory

A growing body of research confirms that a newborn whose cries are left unattended suffers "toxic" stress. If this continues over time, that child's brain connections are permanently, negatively impacted. The result is a child who grows up with more difficulty learning, controlling emotions and behavior, and trusting others. These children also grow up more prone to problems such as obesity, diabetes, cardiovascular disease, and autoimmune disorders.

In direct contrast are children whose parents Tune In during the first years of their lives, responding to them promptly and positively. In addition to building their children's brains, these parents help lay the foundation for what researchers refer to as *attachment*. Documented across cultures, the concept of attachment provides an explanation for how the parent-child relationship develops, ultimately shaping the child's social-emotional and cognitive development.

The theory of attachment was first hypothesized, in 1951, by British psychologist John Bowlby, whose work with emotionally disturbed children led him to take a closer look at the impact of a child's relationship with its mother on social, emotional, and cognitive development. Based also on the evolutionary theory of survival, that is the importance of a child's maternal protection against predators, the theory has been altered somewhat since Bowlby's original insights, but the importance of the maternal or primary caregiver relationship to the emotional development of a baby and young child has a growing body of research to support it.

The Many Colors Of Communication

Before they have real words, young children communicate in different ways. Newborns cry. How else will you know if they are hungry, tired, bored, or lonely? As they get older, they coo, they gurgle, they babble, they point or make silly faces responding to an adult's amused attention. As newborn reflexes become better controlled, they may try to get a parent's attention by arching their backs, kicking, or squirming. We're pretty sure that these behaviors are done for the parent because there is usually an attempt at eye contact, as well.

Now think about how really smart a baby is, only recently out of the womb and already devising effective ways to get a parent's attention. The pushing, smiling, gurgling, pouting, may make a baby appear cute, but the cuteness is really only a ploy hiding a very clever, very effective, very real language to get something he or she wants.

Now think how smart parents are, too. Because their first frantic efforts in parenting are to learn that language really well.

And it isn't an easy language to learn. What are the communication clues in a gurgle, a wail, or an assortment of other sounds? Until words become integral to a child, decoding this language is easier said than done, requiring time and a lot of trial and error. Even with concerted effort, it's often impossible to know for sure. But making the effort is important. In addition to providing a baby with a greater sense of security, it is another important factor in establishing the parent-child relationship, a key factor in optimum brain development. It is the basis of Tuning In.

The Second T
Talk More

The second T, Talk More, refers to more than just numbers of words; it's the kinds of words and how those words are said that are the salient factors.

Imagine the brain as a piggy bank. If all you stick in are pennies, even a filled piggy bank won't do much toward paying tuition for college, let alone medical school.

In the same way, if the only words you put into a baby's brain are three-for-a-penny words, there won't be much to put toward college tuition, either.

On the other hand, if you put in a wonderfully diverse vocabulary, day after day, and that brain becomes really rich, it may be able to pay its own tuition.

Talk More, which goes hand in hand with Tune In, refers to a parent's increased talking *with* a child, especially about what the child is focusing on, not *to* him or her. While this may seem a subtle distinction, it is fundamental to the TMW approach. Talking More *with* a child requires a mutual level of engagement between the child and the parent. Like Tune In, it is another critical element of parent-child attachment and brain development.

Narration

Narrating what you are doing, while you are doing it, can sound pretty crazy to someone listening to you. But narration, another method of surrounding a child with language, in addition to increasing vocabulary, shows the relationship between a sound, that is, a word, and the act or thing it pertains to. Wash. Dry.

Diaper. Hand. The routines parents take for granted are valuable to the young child; every word, every description, transforming otherwise ordinary events into brain building and attachment building.

"Let Mommy take off your diaper. Oh, so wet. And smell it. So stinky!"

"Now we can put on a new diaper."

"Mmm, look at this new diaper. It's white on the outside and blue on the inside."

"And it's not wet. Feel. It's dry and so soft."

"Isn't that much better?"

"Let's put your pretty pink pants back on."

"Wet or dry, Mommy loves you!"

Narration is also a way to familiarize a young child with the steps involved in routine activities. Although the parent will be doing most of the work, the goal is for the child to eventually do it alone.

"It's time to brush your teeth. What do we do first?"

"Get your toothbrush! Your toothbrush is purple and Daddy's is green."

"Now let's squeeze the toothpaste onto the bristles."

"A little, little squeeze. Good job."

"Now we are ready to brush, brush, brush. Up and down, back and forth. Let's brush your tongue. Ooh, doesn't that tickle?"

Through this process, parents are building vocabulary, cultivating independence and, as an added bonus, saving on future dental bills.

Parallel Talk

Another aspect of Talk More is parallel talk. While narration occurs when parents talk about what *they're* doing, parallel talk is commentary on what the child is doing. Tuning In is a strong component of parallel talk.

"You have Mommy's purse."

"The purse is so heavy."

"Should we see what's inside?"

"Ah, you found Mommy's keys."

"Not in your mouth, please. We don't chew on keys.
     They're not food."

"Are you trying to open your truck with the keys?"

"The keys open the door."

"C'mon. Let's go open the door with the keys."

Both narration and parallel talk are strategies that can be used from birth. These strategies do have qualifications, however. They should never be laden with repetitive questions or long, complicated sentences. At their best, they include establishing eye contact, talking about things in the immediate environment and, when possible, holding a child close, allowing the child to absorb both language and warmth.

Take "It" Away

Pronouns, to adults, are like air: essential, yet you can't see them except in your mind, and then only if you know what they refer to. He . . . she . . . it? Your child will have no idea what you're talking about. Uncle Michael, Grandma, the sink? Ah, now I see! This is true not only for a child. What if I were to ask you, "Can you please go there and pick it up?" Where would you go? What would you get? For the same reason, labels, "house," "car," "road," "pizza," are very important for both vocabulary building and understanding for the young child.

Your child gives you his scribbly work of art. How do you feel about it?

"I love it!"
No, you don't.
"I love your drawing!"
Yes, that you do!

Every label is another word, another bit of understanding and another boost to a child's brain.

The beautiful thing about these simple techniques is that they are applicable to children of any age and any vocabulary. The more a child is surrounded with rich language, the more familiar a child becomes with hearing words and learning their meanings, the more he or she will be comfortable using them.

Decontextualized Language
A Fancy Term For Not Talking
About The Here And Now

When children first start talking, their language typically conveys information about the here and now. Objects they see are labeled, "doggy," "boo-boo," "Mama," or describe events they are participating in, "fall down," "go potty," "no go sleep."

These words, referring to a visible object or action, are referred to as "contextualized language." As children get older, however, usually between the ages of three and five, they begin to use language about things or events that they are not currently seeing or experiencing. The term for this is "decontextualized language."

Advancing to this level of language is an important sign of intellectual progress. While contextualized language focuses on things or actions that are in view, and benefits from gestures, facial expressions, and intonations to communicate the meanings of words, decontextualized language has no such support. Relying almost entirely on already-learned words, but with no observable reference, it takes a higher level of thinking for processing and responding. Not surprisingly, it is felt to have a significant relationship to a child's brain development.

Using decontextualized language when Talking More with a child is not difficult. It simply entails using familiar words to talk about things that a child and the parent have done together, about a toy recently played with, or about someone he or she knows. The child then has to tap into existing vocabulary to understand, without the support of clues from the immediate environment. Being able to understand and respond to decontextualized language optimizes school learning since so much of academics involves decontextualized language without the advantage of a parent standing by to explain.

Expansion, Extension, Scaffolding

Charades are a good way to characterize early child-to-parent communications. How does a child tell a parent he or she wants to be picked up? By raising both arms. Even when words are used, they are usually basic and concise: "Sit." "Milk." "No." Learning a language for the infant and young child is not a passive event. While we are all born with the capacity, the development of a complex language structure is absolutely environment dependent. A child who regularly hears appropriate and meaningful language will eventually use that language.

"Uppie, uppie."

"You want Daddy to pick you up?"

This exchange, over time, will evolve to:

"Please pick me up, Daddy. I'm tired."

A child learning to talk uses partial words and incomplete sentences. Language expansion, within the context of Talk More, restates what a child is saying by filling in the blanks. The expansion of:

"Doggy sad."

is:

"Your doggy is sad."

Language expansion smoothly offers the child a better way of saying something without the negative aspect of "correcting."

As the child grows older, language extension adds more complexity.

"Go night-night."

becomes:

"You want to go to sleep. It's very late and you're so tired."

Language extension uses words a child already knows as building blocks for more elaborate communication. This may include adding a verb, an adjective, or a prepositional phrase.

"The ice cream is good."

becomes:

"This strawberry ice cream tastes so good, but it is so cold!"

Scaffolding helps build language skills by adding words onto a child's response. For example, when a child uses one word, parents respond with two or three; for a child who uses two or three words, parents use short sentences.

Expansion, extension, and scaffolding are all methods of staying one or two steps ahead of the child's ability to communicate, encouraging more elaborate, detailed communication, an important goal of Talk More.

The Third T
Take Turns

The final T, Take Turns, entails engaging a child in a conversational exchange. The gold standard of parent-child interaction, it is the most valuable of the Three Ts when it comes to developing a child's brain. In order for the necessary serve-and-return of conversational interaction to be successful, there has to be active engagement between the parent and child. How does the parent achieve this? By Tuning In to what the child is focused on and Talking More about it. The key, whether a parent has initiated interaction or is responding to a child's initiative, is for the parent to wait for the child to respond. That is what sets the stage for the critical Taking Turns.

How a parent Takes Turns with a child will change quite a bit as the child grows. Babies, even before they can talk, can be effective communicators. A crying infant is communicating that a diaper needs changing. A baby who rubs its eyes is saying it's time to go to sleep. Conversation with a baby means reading communication clues, decoding what those clues mean, and responding. It may not be what's considered a typical conversation but these back-and-forth exchanges are important for building both a baby's brain and the parent-child attachment.

As the baby becomes a toddler, Taking Turns becomes more varied. For the child, facial expressions and gestures carried over from infancy become embellished with made-up words, an approximation of words, and real words. A parent's responding to these signals, then waiting for the child's response, is particularly important now. Emergent talkers often have to search for words. It may take so long that a parent's instinct may be to respond *for* the child. This may expose the child to more language, but it may also end the conversation. Allowing the child a little extra time to retrieve words can be the difference between continuing Taking Turns and ending it.

One word that has a limiting effect on Taking Turns is "What?" "*What* color is the ball?" "*What* does a cow say?" "What" questions are low on the totem pole for enhancing conversational exchange or building vocabulary because they only ask a child to retrieve words he or she is already familiar with. Questions that are answered with a yes or no fall into the same category, doing little to keep a conversation going or teach the child anything new.

Open-ended questions, on the other hand, are the way to go, meshing wonderfully with the goals of Taking Turns. Particularly for a young child, they are great conversation starters and conversation continuers. A simple "how" or "why" allows a child to respond with a wide range of words, thoughts, and ideas. There is no way to answer a why question with a nodded head or a pointed finger. "How?" and "Why?" start a thinking process that can lead, eventually, to the skill of problem solving.

THE THREE Ts AND TECHNOLOGY MAKING SURE WE'RE GETTING IT RIGHT

We've discussed the negative aspects of digital technology, especially when it acts as a wedge between the parent and child while emails are written and iPhones are answered and the what's-happening-in-the-news flash is read. But digital technology has been an important factor in helping TMW get it right.

The LENA

The LENA, or Language Environment Analysis System, offers an important window into a child's early language environment. Essentially a word pedometer, the LENA is a small digital audio recorder that fits snugly into the pocket of a child's T-shirt especially designed for it. While being worn, the LENA records the sound environment for up to sixteen hours. The resulting digital audio file is then uploaded to a computer for ongoing comparison with the baseline recordings done before and those done after the intervention began.

The LENA was developed by Terry Paul, a successful entrepreneur. With his wife, Judi Paul, Terry Paul had previously started Renaissance Learning, a technology-based company for developing math and literacy skills. Despite its success, reputedly he had felt that its impact on children came too late. Legend has it that when he read Hart and Risley's *Meaningful Differences* he immediately knew he wanted to develop a technology that would measure a child's early language environment. His favorite mantra was "If you can't measure it, you can't change it!"

Just as the pedometer has been shown to encourage sustained physical exercise, the LENA, in addition to supplying feedback on a child's language environment for researchers, has become an important tool for encouraging improvement in a child's language environment. By offering parents a mechanism by which they can set, track, and assess attainment of goals, it offers encouragement when efforts fall short or confirms improvement when goals are met or exceeded. It's an important tool for motivation.

We first used the LENA at TMW to see if our initial curriculum helped increase the amount parents spoke with their children. While we found that it did, we also discovered that the increase was temporary. The graph lines would shoot up, only to plunge down again: the kind of results that either set you back or set you thinking. We started thinking. Our first thought: are we the only ones who should be looking at the results? This was an easily answered question, leading us to a meeting with our parents to get their thoughts and advice; a giant step forward on the way to developing TMW's well-honed, finely tuned program.

Our Key Motivation

When milk for a baby is not available, there are ways of creating substitutions that keep the baby alive and healthy. Nourishment that allows the human brain to reach its full potential is, however, wholly dependent on language from a warm, responsive adult in the first years of life. Science has shown us that, so far, there are no substitutes. Helping it to happen for all children is what motivates us at TMW, those of us who are the researchers and those of us who are the parents. Our goal is the children.

PART II
THE THREE Ts In Action

As we've said, the credo of the Thirty Million Words initiative is the malleability of a child's brain; its core is the Three Ts, its purpose is to ensure the optimum intellectual development of all children. To this end, the TMW curriculum is designed to enhance the language environment of young children from birth through three years of age. But the effect of the Three Ts goes beyond building vocabulary, including such diverse areas as the introduction of math concepts, developing literacy, building self-regulation and executive function, and developing critical thinking skills, emotional insight, creativity, and persistence. TMW takes science and puts it into brain-developing action.

Book Sharing

Talking with a child from birth lays the foundation for a child's communication skills long before he or she ever talks. In the same way, reading with a child from the first day of life develops literacy skills and a love of books way before the child has the ability to read. As with talking, how and how much a parent reads to a child during the first few years of life has a significant impact on the child's school readiness and ultimate life trajectory.

The importance of reading with a child is not new. Organizations such as Reach Out and Read, Raising Readers, and Reading Rainbow have been advocating its benefits for decades. In 2014, the American Academy of Pediatrics announced its new recommendation that all parents should read to their children from birth.

There is plenty of science to support this philosophy. Studies show that children who have been read to in the earliest years have a larger vocabulary and better math skills when they start kindergarten. There's also evidence that a parent who shows an enthusiasm for reading strengthens a child's interest in learning to read and puts the child on the road to being a more successful reader.

While they were aware of the importance of reading to their children, however, many of the mothers in the TMW initiative didn't like doing it initially.

"He won't sit still."

"She wants to hold the book herself."

"She doesn't let me finish the page before she tries to turn it."

"He's always interrupting me to talk about things that already happened in the book."

We understood what the mothers were saying: that in their minds, the prerequisite for successful reading to a child was a quiet child who listened. Otherwise, there was no sense in reading. What they had to understand, and many of us had to learn, was that that was the perfect time to be Tuning In.

How The Three Ts Help In Book Reading

Traditional story time has the parent reading and the child quietly listening. In "dialogic reading," based on Dr. Grover Whitehurst's Stony Brook Reading and Language Project, the roles shift slightly. The goal in the project is to encourage children to take a more active role in telling the story, including asking questions and talking about what they see, think, and feel. When that happens, the child becomes the storyteller and the parent becomes more of an audience. The TMW approach, referred to as book sharing, is rooted in this approach.

For example:

A children's book is on the parent's lap opened to the first page. Traditionally this means starting to read the book, beginning to end. But the Three Ts use a somewhat different process. Parents, while reading, remain acutely sensitive to which parts suddenly grab the child's attention, adjusting their own focus accordingly. In other words, Tuning In. As a result, the child has an open, clear pathway to learning because nothing is forcing attention to something in which he or she is not interested.

Talking More is the second component in book sharing. The advantage of Talking More for brain building is easily understood. While the level of detail will change as the child gets older, Talking More about what's going on in the story, what this might lead to, how it affects the characters, gives the story greater meaning in the child's mind. In addition, while books use everyday, familiar words, they are also filled with rich, complex, and rarely used words, like "scamper," "mischievous," and "magical." Repeating these words in conversation about the book helps solidify them in the child's mind.

"Baby Bear is sitting at the table."

"Look at the steam rising from his porridge. It's piping hot! What do you think will happen if he eats it now?" "Maybe he should wait till it cools!"

"Oh, no! Goldilocks sat in Baby Bear's chair! What happened to it? The chair crumbled into a pile of pieces. What a mess!"

For slightly older children, Talking More can include Taking Turns with open-ended questions about events, thoughts, and feelings as they relate to the story. Requiring more reflection, more what-if conjecturing, these questions require a higher level of imaginative thought on the part of the child since the answers may not be taken easily from the page. They are excellent opportunities for using decontextualized language.

> "What happened when Goldilocks sat in Baby Bear's chair?"
> "Should she have done that? Why not?"
> "What do you think will happen when the Bear family gets back?"
> "What's Baby Bear going to do when he sees that his chair is broken?"
> "What do you think the Bear family would say to Goldilocks if they saw her?"

Taking Turns, another aspect of book sharing, occurs every time a baby or child points to a picture, opens a flap, turns the page, asks or responds to a question.

TMW's book sharing is not, of course, meant to stop parents from actually reading books to their children. If a toddler wants to climb on a parent's lap and listen quietly to a story, by all means, that's what the parent should do. There are few things more wonderful than snuggling with your child and reading a book. In fact, if that's what a child wants, it's a wonderful way to Tune In.

Reading To A Baby

The American Academy of Pediatrics agrees with TMW that, with only a few simple modifications, book sharing should include infants. While babies will not understand the words, they are comforted by the sound of a parent's voice, the rhythm of the speech, and the warmth of the touch. And while the first enticement of hearing a book may be the nurturing voice of a loving parent, the cadence of words strung together into sentences is a very early lesson in how language works.

Because comprehension is not the goal when reading to a newborn, it is not necessary to choose a children's book for the activity. This may be a good time, in fact, to catch up on the news or finally crack open the bestseller that's been sitting on your nightstand for the last six months. Just turn to the first page and start reading aloud.

A baby begins to show interest in books at about four months of age, although the focus may be the physical book rather than listening to the story. The role of the parent is to Tune In to what is attracting the baby's attention and Talk More about it.

"You want to hold the book so you can get a better look at the pictures. That's a dog. And what is that? A cat, right?"

"Listen to the noise your hands make when you pat the pages. That noise makes you smile. Mommy is going to pat the pages, too. Now Mommy is smiling."

"You think dropping the book on the floor is funny. Look how Daddy has to bend to pick it up. It is funny, isn't it? Let's do it again!"

Print Awareness

A lot of research supports reading with a child to build vocabulary. But there is another factor essential to increasing a child's ultimate ability as a reader: print awareness.

To a toddler, letters are a jumble of lines with no apparent meaning. Learning to read is dependent on understanding that those lines are sound-producing letters that, put together, form words. Gesturing plays an important role in learning this. When a parent points to the words as they're read, a toddler begins to understand that there is a connection between the word being spoken and the specific lines on the page. This also shows the child the language-specific procedure for reading, in English, for example, left to right, top to bottom, and that individual words are separated by spaces and punctuation. When a child is older and there is an unfamiliar word in a book, gesturing toward the written word is another method of teaching the child that there is a connection between what is being said and the words on the page. This process also helps a child understand the relationship between the text and the pictures. In addition to being a very preliminary stage for eventual reading, it is also an example of teaching print awareness.

The significant benefit of teaching print awareness has been shown in studies. Children receiving the most pointing to text had increased print awareness and demonstrated higher reading, spelling, and comprehension skills when compared to children who had been read to without gesturing.

Storytelling And Narratives

Many of the vocabulary and preliteracy benefits a child receives from books are also present in oral narratives, or storytelling. Research has demonstrated a clear link between parents' oral narrative activities and their children's later language skills and readiness for school. Three- and four-year-old children of parents who had received training in oral narrative, that is, storytelling, showed significant improvement in their decontextualized vocabulary, demonstrating that parents' use of narrative helps shape future vocabulary in their children.

Storytelling is not just about reading books. Nor is it only about imaginary kingdoms, beautiful princesses, or dogs floating in outer space. That's one kind, of course, but for a young child, storytelling can just be what happened on a recent trip to the grocery store, the walk in the park, a ride into town, or the bubbles in a bath. While these may seem like rather dull plot lines, children love it when they're the stars!

Again, the Three Ts are great supports. When stories involve shared experiences, not only are they relatable to a child but they also encourage participation. Talking More about those experiences helps children Tune In and Take Turns by stimulating them to add information and additional thoughts. This can be encouraged via open-ended questions such as "What do you think happened next?" "Where do you think they went?" "Why do you think they did that?" This kind of storytelling encourages imagination, vocabulary growth, and in-depth thinking.

As a child grows, so will his or her involvement in storytelling. While a parent, in a sense, narrates a story to a baby, once a child is old enough to participate, daily storytelling can become another factor in brain building. Older children may Take Turns by elaborating on a story or adding the next segment in a saga about a trip to Aunt Susie's. Stories can eventually become more personalized with in-depth questions, including those about "thoughts and feelings" related to the topic. In that way, Tuning In, Talking More, and Taking Turns become key for encouraging active, interested participation.

Storytelling can also help a young child make sense of "feelings." A toddler who tumbles off a slide may be afraid to play on it again. A child who loses a beloved stuffed animal may be very, very sad but unable to express it. Using storytelling as a tool to describe an event and the emotions surrounding it helps a child, as it does an adult, to better understand what happened and to begin the process of easing the anguish. Doing this consistently and appropriately will help a child learn how to understand, identify, and express emotions and even to develop better self-regulation.

Math And The Three Ts

The Thirty Million Words math module has been extremely well received by our parents. Its strategies, designed to help parents harness the power of their words to build a child's early math foundation, are easy to put into practice. They are so simple, in fact, that many parents discovered they were already using them. Perhaps the most surprising realization was that *talking* about math to young children lays the groundwork for greater math ability when they enter school.

The topics that are central to building a strong early math foundation are numbers, their operation, geometry, spatial reasoning, measurement, and data. The basics for each of these are learned, very subtly, very early in life.

A baby who fusses when a stranger picks it up is using math: making comparisons, correlations, and distinctions. Think of it like this:

> Familiar smell = Good
> Unfamiliar smell = Bad

This requires the mathematical skill of collecting and organizing information. Later in life, it will evolve into the ability to sort and classify, helping the child think logically and make rational sense of the world.

A toddler who eagerly requests *more* ice cream is using comparative measurement, another math concept.

A three-year-old jumping in to sing "E-I-E-I-O" at exactly the right time while singing "Old MacDonald" is using patterns, another concept central to mathematics. Being able to recognize patterns helps a child develop problem-solving skills and the ability to make predictions.

The most obvious place to start building a child's early math foundation is, of course, with numbers and learning to count. Initially, learning to count is done via rote memorization: one, two, three, four, with no comprehension that these words represent total quantities or that their relative positions relate to those quantities, in other words, that ten is more than six or two just because it comes *after* six and two when we recite. Over time, however, the knowledge that a number represents the total quantity of a particular set, that is, that the number four represents the total number of cookies on the plate, will be understood. This concept, called cardinality, as we've already discussed, is essential for development of later math skills.

This is because numbers not only add, they also represent a total quantity of individual elements, a relative position among other components, and measurements. They are even used as identifiers. In order to be successful in math, a child must come to understand how numbers work in each of these contexts. The Three Ts offer strong support for easing the way in this very complex process.

Numbers, Numbers Everywhere
And How They Make You Think

Numbers are everywhere. On envelopes, inside shoes, on the television remote control. The more children see numbers, the more they're pointed out, the sooner they will be able to identify them on their own.

Count a baby's toes during a diaper change. Count each piece of cheese on a toddler's plate, pointing to each piece while counting. Ask a preschooler to count the steps as he or she climbs the stairs. When a child is older, start with the total number of objects, then point to each and count. "There are ten toy cars: one, two, three, four . . ." This teaches cardinality, that each object is counted only once and that the number refers to the total of things in a "set."

All that's needed to make mealtime, playtime, just about anytime, an enjoyable number-learning experience for a child are the Three Ts and something to count.

> **Tune In:** A parent notices that a toddler wants to help dress him- or herself in the morning.
>
> **Talk More:** "Your romper has five snaps. Can you help Mommy count them? One, two, three, four, five. Five snaps to snap and you'll be ready to go."
>
> **Take Turns:** The child takes turns by snapping the snaps and counting with Mom. One . . . Two . . . Three . . .

For the older child, counting can include simple addition or subtraction.

"You have two crackers and Mommy has two crackers. Together we have four crackers."

"But what if Mommy gives you one of her crackers? Then you will have three crackers and Mommy will only have one."

This simple strategy adds another math concept for the child.

Geometry

Believe it or not, for young children geometry is really fun. Because geometry means building a tower with wooden blocks, working on a puzzle, or tossing brightly colored beanbags into a basket. Best of all, the parts that create this fun experience called geometry, including manipulating shapes, spaces, and locations, are fundamental to building a strong early math foundation.

Using the Three Ts to talk about shapes and their relationship to each other is a great way to start. Children are already surrounded by the best teaching examples possible. The kitchen door is a rectangle. The dinner plate is a circle. The picture frame is a square. The tile is a triangle.

Then there are the shapes within shapes. The pillow is a square but its cover is a swirl of polka dots. The refrigerator is a tall rectangle with two smaller rectangular doors. Every bit of life with a child is an opportunity to explore numbers and shapes: the park bench, a double-decker bus, the cans on a supermarket shelf, an ice-cream cone.

Integral to learning geometry is something we've already described, spatial reasoning, or learning about how shapes relate to one another. Spatial reasoning refers to the ability to visualize shapes or objects in different positions, mentally "manipulating" them, imagining their movement in relation to one another. We use spatial reasoning when we tie our shoelaces, pack leftovers in a plastic container, or merge into freeway traffic. A young child uses it to do a puzzle, put away toys, or climb onto playground equipment.

Spatial words include the names of the shapes themselves, like "rectangle" and "square," and words that describe the shapes, like "curvy," "straight," "tall," "shorter," and "zigzag."

This represents another instance of the importance of language. Susan Levine's study, as we've noted, found that children who knew more spatial words at age two had better spatial skills at age four and a half.

Spatial reasoning has also been shown to be a key predictor of surgical ability. A surgeon entering an operating room is most likely also mentally entering the anatomy of the body and envisioning the specific pathways necessary to completing a successful operation. It's interesting to think that it might have all begun with a puzzle at three years of age, or younger.

Even if a child does not decide to become a surgeon, the process of configuring a puzzle piece into a puzzle or building a fort out of blocks or replacing a book in a bookcase is an important factor in spatial reasoning. Studies have shown that mastering spatial reasoning bolsters problem-solving skills overall, serving as an important predictor of reading skills and later achievement in science, technology, engineering, and math fields. While spatial thinking can continue to be honed into adulthood, starting early is an important first step in strengthening a child's math foundation.

The Three Ts And Spatial Ability

Applying the Three Ts to spatial talk is an effective way to cultivate a child's spatial ability. Look for opportunities that are a good fit for spatial talk, including those using words that denote dimensions, such as "big" versus "little"; shapes, such as "square" versus "circle"; and spatial properties, such as "flat" versus "curvy." Playtime activities like building blocks, drawing pictures, doing age-appropriate puzzles, and routine activities such as making the bed or putting away toys are all great opportunities to use spatial talk.

Bath time is also a perfect time to use the Three Ts to build a child's spatial ability.

> **Tune In:** A toddler loves the bubbles in the tub.
>
> **Talk More:** "The bubbles are like a big white blanket. And you have a bubble line on your arm. It's a straight line. And, look. I found a little round bubble island. It's surrounded by water. The bubble island is close to your hand. But it's far away from your toes. It's a circle. Can you make other circles in the water? Can you make a square? That's very hard. How about a high mountain?"
>
> **Take Turns:** "The bubbles are covering your hands. There are a lot of bubbles, aren't there? What shape are the bubbles? You're right, they're round! And look at the soap floating in the bubbles. What shape is the soap? A rectangle, right? And your washcloth is square. Let's put the soap in the washcloth. Now we have a rectangle in a square!"

The reward for all of this will undoubtedly come in the future, when a child's math and spatial reasoning ability evolve into skills that open the door to a lot of exciting careers.

Measurement

Measurement is so integral to our lives that teaching its basics early makes good sense. Measurement is used to cook, to clean, to know how many steps to take, to know how much to put on the dinner plate. When we build a shelf, slam-dunk a basketball, or decide how much money to put in the parking meter, we are measuring.

Language, coupled with concrete experiences, is a child's first introduction to measuring.

"Can you make your choo-choo train go really fast?"
"Wow! The tower you built is very tall."
"This box is so heavy I can't lift it."
"That piece of spaghetti is really long."

After a child has developed a sense of attributes such as length, weight, height, and speed, measurement through the lens of comparison can be learned.

"Which of your choo-choo trains goes faster?"
"Wow! Your tower is taller than the lamp."
"Maybe I should pick up the smaller box. This one is so heavy I can't lift it."
"That piece of spaghetti is longer than the plate."

And:

"You've grown so big. Now your shirt with the monkeys is too small. You need a bigger shirt!"

"Your cup was full before breakfast; now it's empty.
    You drank it all."

"Look how far you threw your ball! I didn't throw
    mine as far. See how much closer it is?"

"If you help me, we can make a cake. Here's a cup.
    Can you fill it to the top with flour? Great. Now
    we'll need sugar. We need less sugar than flour.
    One-half cup. Can you fill the one-half cup to
    the top? Great again. I love baking with you."

Comparison words such as "big," "small," "full," and "empty" also help a child to understand the comparative concepts of "same" and "different" and "more" and "less."

A Kid's Guide To Collecting And Understanding Data

While understanding data may seem to have little practical application for children, it's actually already a part of their lives and, as such, another important component of a comprehensive early math foundation. In order to understand the world, children are wired to notice and collect information, that is, data, about the people they encounter, animals, the weather, things in a room, the taste of macaroni, in other words, absolutely everything. It's how they learn to make sense of the world they live in, and their place in it.

Early data collection and analysis happens when a baby is fed a new food, makes a face, and spits it out. It occurs when a toddler has to choose between two cookies of different sizes. It happens when a little sister separates the orange gummy fruits from the green gummy fruits, and gives her brother the smaller pile. It happens when a preschooler looks at the size of his toy truck and compares it to the size truck his buddy has.

> **Tune In:** A child walks around the living room wearing his father's shoes.
>
> **Talk More:** "You're wearing Daddy's shoes. They're sure big on you! Daddy has big feet so he needs big shoes. Look at the difference in Daddy's feet compared to yours. Yours are much smaller."
>
> **Take Turns:** "Whose shoes are bigger? Daddy's or yours? Right! Daddy's shoes are much bigger than yours. But your feet are growing. That's why we needed to buy you new shoes last week. Your old shoes were squeezing your toes. They were much too small."

Patterns

Recognizing the nuances, that is the subtle variations, in sets of information, or data, that we see every day is part of learning to understand how things can fit into a pattern. Being able to recognize, identify, and create patterns helps a child think logically and make predictions, skills that are essential not only for learning math, but also for making sense of everyday life. Patterns help a child learn to count, to read, to play music, to tell time.

Adults use patterns all the time. In order to design a sales strategy, a business owner looks at patterns in sales. To program software, an information technology specialist uses patterns in code. A trash collector uses directional patterns to follow his route. A doctor uses patterns in health to diagnose illness.

Children use patterns in much the same way. A baby can predict that after a diaper is changed, Dad is going to replace the sleeper. A toddler can predict that after lunch comes naptime. A preschooler can predict that right after Mom and Dad come home from work they'll all eat dinner. These events are predictable because everyone, even very young babies, recognizes patterns in their daily lives. It is the familiarity of patterns, in fact, that provides children with much of the comfort they find in routines. When children know what happens next, their brains can focus on learning.

The Three Ts can help teach a child about patterns. Babies like repeating the sounds they hear, so when a baby babbles, keep the serve-and-return going as long as possible. Take advantage of a toddler's love of singing and dancing. Sing a song with a familiar, repeated refrain, especially one with a bouncy dance attached, and encourage participation. Visit the park with a preschooler and Take Turns finding patterns in the playground equipment or landscape. Finding patterns can be done in the laundry, at the dinner table, at the zoo, on the sidewalk, or in the car. Patterns are everywhere, as are the opportunities to talk about them.

Finally, math may be one of those foundational skills. Dr. Deborah Stipek, professor at the Stanford University Graduate School of Education, has written, "Studies [have shown] that children's math skills when they enter school are very strong predictors of their academic success later on. One study showed that math skills upon kindergarten entry predicted children's reading abilities in third grade as well as their reading skills at kindergarten entry. While children can learn beginning math skills after they enter kindergarten, they will be at a disadvantage."

Conversely, a child who starts school with the seeds of math understanding already planted will have a jump start on learning.

TMW
AND
PROCESS-BASED PRAISE

What we want for our children is clear: the ability to reach their potentials, stability, productivity, empathy, constructiveness, and, of course, persistence in the face of obstacles. What is the difference between the child who tries and tries again and the one who tries, fails, and stops?

As we've already discussed: Praise.

When some TMW parents expressed concern that praising too much would give their children a "big head," we helped them understand that their children were looking to them for reinforcement, support for what they were doing and how well they were doing it. But, like our mothers, most of us had to learn that not all praise achieves the best results. Repeating what we've learned from the work of Professor Carol Dweck, there are actually two types of praise:

> **Person-based praise praising the child:** "You are so smart."
>
> **Process-based praise praising the child's effort:** "You worked hard on that puzzle and you finished it. Great job!"

Research shows that children who hear more process-based praise, who are praised for their effort, are less likely to give up when faced with a challenge, a persistence that will help them do better in school and in life.

Consider a young child doing a puzzle. Mom joins the activity on the floor and Tunes In. The child tries a piece in multiple holes before he or she finds the right one. Mom responds with a process-based encouragement.

"I love how you kept trying until you found the right spot for the puzzle piece. You were determined! And you did it!"

The child begins to learn that not giving up is a strength.

How is more process-based praise incorporated into daily interactions with a child? Look for opportunities to *catch your child being good*. Remember, a young child is really still learning the basics of "good behavior." Pointing it out whenever possible reinforces what it looks like. Tuning In is essential here. If attention isn't focused on the child, those ordinary moments when the child is doing exactly the right thing will slip away unnoticed, whereas the wrong things will always be obvious and criticized. Praising "good" encourages its habit.

"What a nice job you're doing sitting at the table while you eat. Daddy is so proud of you."

"You're really concentrating on your drawing. I love all the colors you're using."

"Kitty loves it when you give him gentle touches. He's purring because it feels so good."

The more specific and consistent the praise, the easier it will be for a child to understand, and more important, to learn what good behavior is.

TMW
AND
SELF-REGULATION AND
EXECUTIVE FUNCTION

Intelligence is important, but if a child can't sit still, follow directions, or control emotions, learning will simply not occur no matter how smart the child is. Executive function is yet another example of the importance of the early language environment. Not only does caregiver language help build children's brains; it can also help shape their behavior.

Executive function was not part of the original TMW curriculum. Its inclusion is an example of how completely TMW mothers are integrated into the TMW program, influencing its shape and development. While the mothers had absolutely embraced TMW strategies for enriching their children's early language environment, they wanted something more: a way to help their children behave better.

Their recommendation was insightful because, in fact, it takes more than being smart to do well in school. Children may be able to count to fifty, sing the ABC's, and even read basic words, but if they can't sit still, follow directions, or control their emotions, they will not be ready to learn on the first day of kindergarten. Without strong executive function and self-regulation, intelligence is often fighting an exhausting uphill battle.

How can a parent help a child develop executive function and self-regulation?

Words.

Just as words can build children's brains, they can also help shape behavior.

We all want to do things we would never actually do. It may be telling the rude so-and-so behind the counter what you really think, finishing off the decadent chocolate cake in the refrigerator, gesticulating to the guy who cuts you off on the highway. It's part of being human. But when those situations arise, we're usually able to control our emotions and resist those impulses. The difference between carrying out disintegrative impulses and calming ourselves into positive behaviors is called self-regulation.

If self-regulation were as innate to humans as breathing, this would be a very different world. What makes one person more likely to control destructive impulses and others not? One reason that we've already discussed is the effect of constant stress in the home. Its effect on the cortisol levels in babies and young children is an important reason for a child's inability to self-control. But even without stress in the home, self-regulation is still a learning process. For that, language is, again, important.

In the early years, before children are able to self-regulate, a parent regulates for them, giving back the toy swiped from a friend, keeping a child from reacting physically out of anger at a sibling, or stopping a child from finger painting on the living room wall. But instilling the ability of self-regulation in the young child has vital, lifelong importance as the critical factor in a child's being able to pay attention, follow directions, problem solve, resist impulses, and control emotions, all of which are crucial for academic achievement from the first day of school. It is another area in which the Three Ts can be used successfully.

It's important to note that the Three Ts are not geared specifically for the development of self-regulation skills, as Tools of the Mind and its sister programs are. But for almost all children, the Three Ts offer parents a routine that successfully supports strengthening self-regulation in all of its facets.

One important way of nurturing self-regulation in children is by offering choices. When all decisions are made by an adult, a child never has to consider actions or the results of those actions. When a child is offered a choice, he or she must stop and consider the options, weigh their importance, make a selection, and then voice or carry out that decision.

> **Tune In:** A child, just out of bed, is eager to visit Grandpa.
>
> **Talk More:** "Let's get dressed for the day. We're going to visit Grandpa later. Here's the purple dress. And there's the pink one. The purple one has such pretty flowers on it. And the pink one has lace on the sleeves. It has pockets, too."
>
> **Take Turns:** "Which one do you want to wear?" "The pink one?" "I thought you'd pick the purple dress!" "Can you tell me why you picked the pink one?" "You wanted it because it has pockets?" "Ah, so you can put Grandpa's candy in it!" "I think it's the best one because the skirt is perfect for spinning round and round." "Very nice choice."

Choices can also work as a way of constructively altering behavior.

Tune In: A toddler is protesting having to sit in a high chair for mealtime.

Talk More: "I think you're hungry and that's why you're a little cranky. Let's have some lunch. Let me look in the cupboard. I see pasta. And pickles. I don't think you want pickles, do you?"

Take Turns: "Would you like a peanut butter sandwich or the pasta?" "Pasta, pasta, pasta! You can't get enough pasta." "Should we put it into a bowl or a plate?" "Listen to the funny noise the pasta makes when Mommy shakes the box. Do you want to shake it? Shake, shake, shake."

Offering choices encourages a child to think independently. Backed up by the Three Ts, it's a great workout for the regulatory part of a child's brain.

The Best Way Of Teaching Self-Regulation? Exhibiting It

Another way of teaching self-regulation is to exhibit self-regulation. Children learn behavior best by modeling theirs after the adults in their lives. When parents are frustrated or upset, they should talk about it, telling their child in appropriate ways, with an appropriate tone of voice, how they feel and how they're handling it. It's important to remember that this is not intended to be a catharsis for a parent, but a way of teaching a child the most appropriate, most constructive way to respond to problems. The Three Ts work here, too.

Tune In: Mom is headed out the door but just realized she can't find her keys. Mom explains without sounding annoyed or stressed.

Talk More: "I don't believe I lost my keys again. This is the third time this week I've misplaced them. I'm really upset with myself. I'm going to be late for work. Can you help Mommy look for her keys?"

Take Turns: "Do you see the keys under the table? That was good thinking to look there because Mommy sometimes leaves her keys on top. They could have fallen. Should we look on the kitchen counter, too?"

This strategy also helps you keep calm when responding to your child.

Tune In: A toddler has dumped his bowl of raisins on the carpet and is walking back and forth, squishing them into the fibers. Dad responds calmly.

Talk More: "Please don't step on the raisins. They'll get the carpet dirty and make your socks super sticky. Let's pick them up and throw them away. These raisins won't taste very good now because they're very dirty. Let's get wet cloths and wipe up the carpet. You take one and I'll take one. We'll do it together."

Take Turns: "You're doing a great job cleaning up from the raisins. Can you take off your socks so you don't make sticky footprints? Great. Now let's go wash our hands and then I'll get you a new snack."

Without question, responding this way takes an enormous amount of brain energy and, for a parent, enormous self-regulation! But by setting a clear example of constructive problem solving, a parent is teaching a child something that will affect a lifetime of handling problems. Self-regulation gives a constructive foundation to just about everything, even child rearing!

Directives Do Not Build Self-Regulation Or Brains

Directives and short commands are the least efficient method for brain building because they require little or no language in response.

"Sit down."
"Be quiet."
"Put on your hat."
"Give me the book."
"Don't do that."

It's really counterintuitive. Telling a kid exactly what you want him or her to do *should* work. And, at the moment, it does seem to. "Stop that!" a parent says in a five-star-general way, and, yes, the child does stop that. "Put on your hat," and, yes, the child puts the hat on. But what has been stopped or accomplished is the action of the moment, not the ongoing habit.

There are countless ways to talk to a child, but not all are equal in building a child's brain. A case in point is directives. Directive language, completely antithetical to the Three Ts, is usually said in an abrasive tone, using harsh language, and seldom requires a response. While words may be present, brain building is definitely not.

The TMW Method

The TMW alternative to directives is called "because thinking."

Daily routines are hectic times. When a young child is added into the equation, taking care of the task at hand can challenge even the most patient of parents. It is those moments of frustration when directives can bubble to the top.

A parent is trying to get a child out the door in the morning: "Go get your shoes."

No thinking required. If the parent is lucky, the child gets his or her shoes.

The TMW alternative:

"It's time to go to Uncle David's. Better put on shoes because otherwise your feet will get really wet from the rain. And they'll be very cold. Please go get your shoes."

"Because thinking" helps a child understand that there is a rationale for doing something, that it's not just a parent-to-child order. "Because thinking" is also a step in learning to gauge cause and effect, the consequences of actions, and why things should be done in a certain way or at a certain time. It's also part of learning critical thinking, an essential for higher learning.

Misbehavior may also provoke an exasperated directive.

A child picks up his parent's phone and pushes on the touch screen with sticky fingers. The response can be:

"Put down my phone! Now!"

Or . . .

"Please put my phone back on the table. If you drop the phone, it could break. Then we wouldn't be able to talk to Aunt Sydnie when she calls to see what we're doing today."

Saying "Eat your breakfast" may get a child to eat. But telling the child *why* sets up a lifetime understanding of the need for food for health.

Saying "Don't play on the stairs" may make the child come down. Explaining *why* sets up a lifetime understanding of the need to evaluate potential danger in certain activities.

This lifetime understanding doesn't happen overnight. But if a parent is consistent, the routine established with "because thinking" will become part of the child's thought processes and the day may actually come when he or she puts on shoes without being told.

As we've already discussed, there are, of course, times when directive language not only makes sense, but is necessary.

A child is chasing after a ball and both are headed into a busy street, directly into the path of oncoming traffic. Not a great time for a placid "Sweetie-pie, please don't run into the intersection. The car that is barreling down the street could hit you and that would hurt."

"STOP NOW! THERE'S A CAR COMING!" is the appropriate directive and, no, it will not help develop the child's brain. But, in that case, it's forgivable.

The positive side is that the critical thinking skills the child is developing because of a parent's general use of "because thinking" will ultimately culminate in a judicious, analytical brain that will say, "Don't!" all by itself. And that, in the end, is what we're aiming for.

CREATIVITY

The Arts, with a capital A, are rarely thought of as a primary focus for children. Yes, there are always crayons and glue sticks, but very often only as a sideline on the way to preparing for medical school. Or engineering. Or learning how to code.

But creativity, even in the sciences, is an important way of discovering new worlds, new ways of doing things, and new ideas that no one else has had. In fact, a child who is encouraged to think creatively when very young will likely have a stronger foundation for learning at the beginning of school. Creativity is not talent or skill; rather, it is the tendency toward exploration, discovery, and imagination. How can a child be encouraged to explore, discover, and imagine? Although the arts are not a formal part of TMW, Three Ts techniques can be very useful here, as well.

The Three Ts And Music

Music benefits a child's brain development on many different levels. It teaches language and communication. It stimulates movement, thereby boosting motor development and physical growth. It builds listening skills. It strengthens neural pathways to the brain that are responsible for abstract thinking, empathy, and mathematics. It provides a creative outlet for expressing thoughts and feelings. It encourages imaginative thinking. The whole child is engaged by music. The whole child benefits from music.

The Three Ts are an easy fit.

Tune In: Singing automatically makes a voice more interesting to listen to, so a child is likely to stay engaged longer.

Talk More: Choose a favorite song and sing, sing, sing.

Take Turns: Every dance move, every clap, every verse shared, is a Turn.

Songs like "Ring Around the Rosie," "Sing a Song of Six-pence," and "I'm a Little Teapot" introduce a child to words not typically said in everyday conversation. Songs like "Five Little Monkeys," "One, Two, Buckle My Shoe," and "This Old Man" introduce a child to numbers and counting. Songs like "Open Shut Them," "The Noble Duke of York," and "Hokey Pokey" reinforce spatial concepts. Songs like "Old MacDonald Had a Farm" and "Bingo" teach a child about patterns. Who knew building a child's brain could be so much fun?

Children love to make music, too, whether it is banging a wooden spoon on a pot, strumming a toy guitar, or pounding the keys of a piano. There are no right or wrong ways to express oneself through music, so it is the perfect occasion to build a child's self-confidence and self-esteem.

The Three Ts And The Visual Arts

The visual arts, including painting, drawing, and sculpting, are also powerful for a child's development. They support motor development and they're also a way for children to express thoughts and feelings they can't verbalize. This is particularly important for very young children who may not have the words to do so. In the visual arts, there is also no right or wrong. It's what pleases the artist. All that's needed is blank paper, crayons, and an imagination. Studies are affirming that children who are involved regularly in art do better in reading and self-regulation as well.

A child's exploration of artistic expression provides a parent with endless opportunities to use the Three Ts.

Tune In: Whatever the activity, whatever the medium, whatever the ideas, follow the child's lead. It may be mixing all the jewel-toned paint colors into one rather boring brown. It may be drawing a series of straight and wavy lines on a piece of paper. It may be covering fingertips with paste and making a path of fingerprints on cardboard. Just go with the artistic flow.

Talk More: Put words to what the child is doing. Talking about artistic endeavors is the perfect time to introduce adjectives and verbs not often used in everyday conversation.

Take Turns: Ask open-ended questions about the materials being used, colors being chosen, what is taking shape, and what else the artist has in mind.

When words are focused on the process of creating, without the need to evaluate or critique, a child is allowed to describe the work in his or her own words. This helps develop the ability to analyze, to communicate ideas, and empowers independence of thought and self-confidence.

The Three Ts And Pretend Play

Pretending is a keystone of child development. A young child who is encouraged to tap into his or her imagination has yet another way to explore the world and, in a sense, begin to add his or own stamp to it. Pretending is a safe portal through which thoughts and feelings can be expressed. It teaches communication and promotes preliteracy skills. It also enhances social skills and promotes higher, more profound, ways of thinking.

Pretend play utilizes a child's existing vocabulary as well as variations of language that have been heard but are not yet entirely understood. An invitation to join in gives parents the opportunity to allow a child to take the "pretend" lead but still learn from a parent's interaction.

> **Tune In:** Be an understudy, a minor character, so that the child is directing. What better place for a child to be in charge than in a world entirely of his or her own creation?

> **Talk More:** Without changing the content of what's unfolding, look for ways to expand and extend the dialogue.
>
> **Take Turns:** Ask open-ended questions to keep the performance going. "What happens next?" "What should I be saying to her?" "What does the castle look like?" "What should I do now?"

A child's imaginative play will change as the child grows. A toddler's pretending tends to be more solitary play, drinking imaginary tea from a toy teacup or holding a wooden block to the ear as if to answer a phone. A preschooler's pretend play starts to be interactive, including role-play and dress-up. Being part of it is another aspect of building brains, relationships, and creative skills, with the added bonus of being fun.

A Final Word

A baby's trying to make a toy squeak is demonstrating creative thinking. A toddler's attempting to build a train out of stacking cups is demonstrating creative thinking. An older child wearing a superhero cape while acting the part is demonstrating creative thinking.

When a child is allowed to express creativity, many, many things occur in the brain. But perhaps the greatest is independence of thought. Math and reading are entirely dependent on learning established rules. The arts are largely exempt from rules. Allowed to flourish, they help a child make sense of the world and to establish a sense of self in it. This may lead, years and years later, to positive, innovative improvements in that world. Another good reason why the arts should be encouraged.

THE FOURTH T

It used to be said that people "zoned out" when they weren't paying attention. Now we can say, with confidence, that they are "digitaled out." It is a disconnect that also contributes to brain development, but not in a positive way. Because there is no Tuning In, no Talking More, no Taking Turns when a parent's responses to a child range from "uh-huh" to "just a second" to complete silence. Maybe the Fourth T should be Turn It Off.

How did parents occupy a child before the digital age? Coloring books? Building blocks? Toy drums? Baby dolls?

And now?

Watch the supermarket aisles as a parent loads groceries into a cart. In it a young, young child is playing on a digital device, usually the parent's iPhone. Don't look at the child. Look at the adults walking by. Do you see any shock? Any surprise? Any "hmmm" reaction in anyone who sees this? Is anyone noticing that there is not a word of interaction between the child and the parent? Does anyone say, "Wow, what a huge missed opportunity?"

Getting the work of life done takes immense time, immense effort. No one is questioning that. But the reason we want to accomplish this mountain of tasks is to make our lives ultimately easier: food in the cupboards, bills paid on time, gas in the car. But a child who is able to learn, who is stable, who can relate to a parent warmly and receptively, also goes a long way to making life easier. In the end, that is a parent's most important goal: a stable child who is able to meet life's challenges constructively and intelligently. Interacting consistently and positively when the child is very young goes a long way to accomplishing that. In the supermarket, a parent can Tune In to what a child is focused on, whether it be the shopping cart or apples . . . Talk More, giving him or her information about that interest and those that follow . . . then Take Turns, planning how to cut the vegetables for the stew or which cereal to buy. Listening to a child is as important as talking, perhaps more so. In fifteen years, a parent who listened as well as talked when the child was young will be very happy with the results. And, yes, life will be easier.

By the way, this thought is not just for supermarkets. It's for restaurants, parks, and bookstores, as well.

TMW is not alone in believing that excessive use of technology is bad for children. The American Academy of Pediatrics recommends no television or technology at all for children under two years of age. For children two and up, their recommendation to parents is that screen time should be limited to less than one to two hours a day with restrictions as far as content. This recommendation includes anything with a screen, including computers, tablets, smart phones, and even electronic games designed for children.

> ~~Tune In:~~ There is no way that a television can Tune In
> to a child. While it may seem as if a child is ut-
> terly mesmerized by what's happening on the

screen, science tells us that no learning will occur. TV is a one-way brain street.

~~Talk More:~~ Shh . . . Ever try to Talk More or, for that
matter, talk at all to someone involved in any
digital device? Can't happen.

~~Take Turns:~~ Digital devices do not Take Turns; they
take absolute concentration. Their part of the
interaction is set; nothing can alter it. Even an-
swering "questions" correctly only means a child
is following orders, not giving and taking.

When shows do integrate questions into their dialogue, the
responses children receive for their answers are preprogrammed,
not Tuned In to the child, nor to the child's reply. The TV cannot continue the conversation. Fun, for sure, but not necessarily
the same quality as parent-to-child interaction.

Dr. Patricia Kuhl's research, described in Chapter 3, studying
two groups of nine-month-old babies who listened to Mandarin,
one via a live person, the other via a person on DVD, showed
that while those watching the DVD appeared more intensely
focused, not only did they learn much less than those spoken to
by a live human, but they did no better than a group who had
heard only English. In fact, the only babies who learned were
the ones who received instruction via real human interaction.

Dr. Kuhl's findings were corroborated by research done at
Georgetown University, this time studying children learning a
novel task rather than the sounds of Mandarin. In this study,
two groups of children, twelve months to twenty-four months
old, watched a person demonstrating, over and over, how to

remove a mitten from a mouse puppet. As in the Kuhl study, one group watched the procedure done by a live person; the other group saw it via a DVD.

The results were essentially the same. The children who saw the demonstration done live were able to imitate the action with little or no difficulty. The children who watched the demonstration on DVD were not able to replicate the action at all.

Conclusion: Children's brains learn best from social interaction.

Dealing With Reality

Is it possible to eliminate technology from any of our lives today? I'm sure you realize I'm typing this on my computer before I send it by email for others to look at, calling them first on my iPhone to make sure they're there and, if they're not, texting them so they know it's coming.

My son, Asher, is downstairs. It's his weekly Friday, school's-over-now-relax-afternoon playing the computer game *Madden 2015* with his six best buddies, Zach, Nolan, Gaurav, Johnny, Jason, and Ben. And, yes, you can tell from the constant shrieks and cheers and advice, lots of advice, that it's the height of interaction. Of course, I prefer when they run outside to play a quick game of touch football but, nonetheless, there is definitely interaction.

So technology does serve a purpose. But it definitely is a habit to take notice of, and when it interferes with a parent's interaction with a young child, it's something to moderate. Thirty Million Words has what it calls a Technology Diet. It involves honestly determining digital intake. In a day, what devices do you use, why, for how long and how much is protein, broccoli, or chocolate? This includes using devices, using social media such as Facebook and Twitter and, of course, Googling to find out what a friend you haven't seen in twenty years is doing. The next step is to see how the devices are interfering with relationships, including a relationship with a child. The final step is programming intake, making a conscious effort to monitor when the devices are used, how they're used, and how much they're used.

What The Future Holds

In the early 1800s, Alexander Graham Bell wrote to his father about something he had invented that would allow "friends to converse with each other without leaving home." Then, on March 10, 1876, in the very first telephone call, he summoned his assistant into his office: "Mr. Watson, come here, I want to see you."

Alexander Graham Bell ushered in an incredibly modern period.

Which should serve to remind us that what is "modern" today will soon be, like a hippy with long hair and a "Make Love Not War" T-shirt, while still a good idea, not modern at all. We are at the tip of the digital iceberg. Tomorrow will be a digitally very different and, very likely, more intrusive day.

It's interesting to note that Alexander Graham Bell refused to have a telephone in his office because he felt it would disturb his scientific work!

Learning To Make Technology A Friend

Lisa Guernsey, director of the Early Education Initiative and the Learning Technologies Project, New America, and Michael Levine, a child development and policy expert and founding director of the Joan Ganz Cooney Center at Sesame Workshop, have thought a lot about how technology might be used to enhance parent-child interaction and children's language and literacy development.

In their book, *Tap, Click, Read: Growing Readers in a World of Screens,* they take us on a journey into the outer space of digital-age learning. Their book examines in rich detail today's century and what it might look like if we developed new ways of thinking and teaching that would help greater numbers of kids. The children whom Guernsey and Levine focus on are those from birth to eight years old.

What Guernsey and Levine wanted to learn was "what it meant to teach children to become literate" in an interactive, digital world, "one in which smartphones, touchscreen tablets, and on-demand video are already nearly ubiquitous." Their questions involve discerning which features and habits related to new technologies will serve the purpose of increasing literacy in young children, which should be avoided, and how answers differed depending on the individual child and different circumstances.

It is encouraging that these questions are being asked. Because there is no way to avoid the increasing presence of digital technology. But while literacy is an essential factor in our lives, the human-to-human interaction that establishes literacy in a young child has wider implications. This is especially true in the early, formative years of babies and young children with their parents and caretakers. The language environment of the child from birth to three years of age does more than affect literacy. It affects the core of who we are. And it depends on more than words; it depends on how the words are spoken, the environment in which they are spoken, and the warm, human receptiveness of the parent or caretaker. It'll take a lot of digital inventiveness to replicate that.

CHAPTER 6

THE SOCIAL
CONSEQUENCES

WHERE THE SCIENCE OF NEUROPLASTICITY
CAN TAKE US

How wonderful it is that nobody need wait a single
moment before starting to improve the world.
—attributed to Anne Frank

What is the ultimate goal of this research? What is the ultimate goal in closing the thirty million word gap? The
ultimate goal of society? It is, of course, to find ways to ensure
that all children reach their potentials, educationally, socially,
personally, productively. This is not only a fundamental philosophy of our country, it is, on a basic level, a way of ensuring our
strength and stability. The science is clear. We all start off pretty
much the same, a bundle of undeveloped possibilities, no matter
the color of our skin, the size of our parents' wallet or the country of our origin. Why, then, is there such striking divergence of
achievement after birth?

When you read this book, it's important not to think it, or

the ongoing research, is about your child or my child or their child, because, ultimately, it really is about the world all our children will have to live in, in the future. It will either be a world of more and more children who will have reached adulthood unable to achieve optimally or a world in which the grand preponderance of its population is educated, productive, and stable, with a great sense of constructive problem solving. Utopia? No. Common, practical, pragmatic sense.

A GROWING PROBLEM

The dramatic increase in income inequality in the United States over the past four decades is mirrored by its effect on our children. In the United States today, more than thirty-two million children, nearly half of all of our children, live in low-income homes. The evidence that this disparity may be accompanied by a widening gap in children's learning outcomes has resulted in a billion public dollars designated for preschool programs. To me, this is both admirable and important. But the hoped-for results are significantly diminished because preschool programs do not impact what research tells us is the precipitating cause of the problem: what happens to these children in the critical years between birth and three years of age. As a result, those billion dollars will be spent largely for remediation of a problem rather than for education.

No Generalities Allowed

It's important to stress that the problem is not simply socioeconomic. Rich or poor, language environments are home and parent specific. This is evident in the current threat to parent-child interaction, no matter the income, in our digital age, whether it is a laptop, iPhone, or iPad. Go to any children's park, watch while the kids are swinging from the jungle gyms, and you'll see what I mean.

In the end, almost all parents, no matter their socioeconomic status or their educational level, have the vocabulary necessary to start their children off on the right path. It's simply a matter of parents understanding the importance of the language environment and, when there is a need, having accessible, readily available supports in place.

If we think of each of our lives as a narrative, an ongoing novel, with ourselves as the main character, then Chapter 1, Page 1 is the preliminary staging for what will follow. We have no control over what happens on Page 1, but as research, including Hart and Risley's, shows us, what is said to us, how it is said, and what is elicited from us will be, in large measure, strong determinants of who we are and how we deal with life; not 100 percent but a sizable percentage, nonetheless, of the rest of the book.

THE DETERMINING FACTORS
PARENTS AND CAREGIVERS

So how does the newborn evolve from his or her initial potential into the realized potential of the adult? That's where we, as parents and caretakers, come in.

While this book on the surface appears to be a story of children and the malleability of intelligence, at its core it's about the essential and powerful role of parents. It's not that parents haven't always realized their importance. Of course we do. Otherwise why would we be worrying about everything we're doing and whether we're doing it right? But rarely, until recent years, has science helped us to better our chances. Not just for our own children but, on a larger scale, by helping us to understand the broader design that could improve the lives of all children and, by extension, the world they will live in.

Recognizing the thirty million word gap, a metaphor for the importance of language in a child's early brain development, is an unprecedented opportunity. It allows parents to understand their power in helping their children realize their ultimate potentials. Even more important, it shows parents the steps to enhance that power. Understanding the thirty million word gap also helps set the stage for turning the tide for all children. In this regard, the science is clear. In order to close the achievement gap, in order to ensure that all children in this country are able to achieve their potentials, well-designed, carefully monitored programs, based on scientific evidence, must exist to help it occur. And those programs, geared at helping the children, are parent/caretaker dependent.

This is what Steven Dow, the executive director of the Community Action Project in Tulsa, calls the *great paradox:* While early childhood is really the story of parents, and although we know the importance of parents in the eventual intellectual outcomes of children, parents are often afterthoughts in program development and reforms for closing the achievement gap. They may be mentioned in the discussion but, in the end, they are usually treated as an add-on rather than the key tool to make the necessary changes. And there's the historical irony. Because it was the failure of their preschool project to help children become school ready that impelled Hart and Risley to do a longitudinal study on parental influences in children's academic outcomes.

SOLVING THE PARADOX

The importance of preschool is not disputed. But when children enter without the prerequisites for learning, it is largely remedial. To give preschool maximum strength, and to make sure that the lack of school readiness does not predict an academic lifetime of "catching up," or failure, the children entering its programs have to be ready to learn. This emphasizes the need to design solid early-childhood programs that include parents to help ensure the school readiness of children who may need additional support. These programs would help parents provide an optimum language environment in the first three years of their child's life, when essential brain development is occurring. Home visiting would help parents set language goals; careful monitoring would help parents achieve those goals. In order to assure success, and to accurately assess program design, programs would include a built-in procedure for evaluation and improvement.

Success will be dependent on a strong support system. While parenting interventions, in the past, have had problems, and may need more research or evidenced-based program development, science demonstrates that making the effort is essential since it will only be when parents, or a child's primary caretakers, are actively involved as engaged partners in a child's early years that outcomes will improve.

It's also true that until we, as a nation, understand the importance of parental involvement, offering appropriate support where such support is needed, the lives of millions of our children will essentially be a lifelong game of catch-up.

Can we really do it?

If we can create a tiny antibody that travels through a body to attack a specific cancer cell, if we can push a few buttons to tell someone in Shanghai that we're seeing a show in Manhattan, if we can send twelve men to the moon, we can do this.

THE CULTURE OF PARENTING

In her landmark book, *Unequal Childhoods,* Annette Lareau, professor of sociology at the University of Pennsylvania, contrasted the parenting styles of different social classes, styles to which she and others attribute the perpetuation of class differences. "In America, social class backgrounds frame and transform individual actions," Professor Laureau wrote. "The life paths we pursue, thus, are neither equal nor freely chosen."

Her findings are the result of research that involved total immersion in the home lives of families with nine- to ten-year-old children from across the socioeconomic spectrum. Her intent was to get a "realistic picture of the day-to-day rhythms of family with elementary school–aged children."

Unlike Hart and Risley, who were simply observers, Lareau and her team said they wanted to be like "the family dog."

"We wanted parents to step over us and ignore us, but allow us to hang out with them."

Rather than collecting numerical data, Lareau and her team used the vehicle of day-to-day sociological narrative of the families to explore whether social patterns had identifiable socioeconomic characteristics. Eighty-eight families were included in Professor Lareau's study. Twelve families were intensively studied with active participation in their lives, including going to baseball games, religious services, family reunions, grocery stores, beauty parlors, barbershops, and even staying overnight.

WHAT DID THEY FIND?
THE SIMILARITIES

Every family, no matter the socioeconomic background or family traditions, had similar hopes for their children.

"All of the families want[ed] their children to be happy and to grow and thrive," said Professor Lareau.

THE DIVIDING POINTS
HOW DIFFERENT FAMILIES
APPROACHED THAT GOAL

Middle-class parents "foster[ed their] children's talents, opinions, and skills" with a kind of frenetic energy. Called "concerted cultivation" by Professor Lareau, they spent hours driving their children to activities, activities, and more activities. There was also "quite a bit more talking in middle-class homes . . . [which probably contributed to the] development of greater verbal agility, larger vocabularies, more comfort with authority figures, and more familiarity with abstract concepts." In addition, parental language in these homes "emphasized reasoning," "verbal jousting," and "word play." Directives were rarely used, "except in matters of health or safety."

Professor Lareau called parenting in lower socioeconomic homes "the accomplishment of 'natural growth.' " The children's lives were much less structured. The only clear absolutes were obedience and respect for authority. Otherwise, it was a much more hands-off approach. Children played freely together, without parental directives, and they developed in a free-flowing way, accepting unconsciously the "ways of the parents" almost by osmosis.

The language of these parents also reflected the difference, with simple directives rather than discussions or debates predominating. For example, a child was instructed to wash by simply saying "bathroom" and being handed a washcloth. While one may analyze why these differences existed, including a stark difference in resources that would allow the time, money, or energy for extraneous words or outside activities, the differences in the children were apparent, especially in educational achievement.

WHY DO ANYTHING MORE IF YOU DON'T KNOW ANYTHING MORE NEEDS TO BE DONE?

Annette Lareau's "concerted cultivation" versus "natural growth" parenting styles reminded me of Carol Dweck's research because, to me, concerted cultivation and growth mindset are similar in many ways. Both imply a belief in a child's intellectual malleability and both employ a deliberate effort to enhance a child's persistence and mastery of skills.

In the same way, "natural growth" has, even if it's unspoken, a sense of "fixed mindset," implying a belief in unchangeable innate abilities. This sense of fixed ability leads, perhaps, to less "concerted" parenting, except, as noted, in emphasizing the role of the parent as the authority.

Is it possible, then, that what we define as differences in the "cultures" of parenting is at some level reflective of the unconscious feelings parents have about the absolutes, or lack of absolutes, in a child's development? In other words, if you aren't aware that you could make a difference in your child's future, why would you do anything differently? As Professor Lareau stressed, all of the parents she followed, no matter their socio-economic status, had similar, positive, goals for their children. It was "how parents enacted their visions" for getting to the goals that was the difference.

This is not to dismiss other impinging factors and the fact that simple "belief" is not the entire story in socioeconomic differences in child outcomes. As Lareau states, the effects of social class are long-term and cumulative, including those involving health care, work opportunities, the criminal justice system, and the political sphere. In fact, understanding the long shadow of social class on life destinations, as well as upward mobility, is an important task for social scientists in our democratic future.

Which got me thinking.

Knowing that parental mindset about the malleability of intelligence had an effect on child rearing, with its eventual effect on the intellectual growth of children, I began to wonder, after reading Annette Laureau's research, if it were possible to see if that mindset was already evident in mothers from the first day of their babies' lives, established even before a baby was born. I looked back at a TMW study conducted at the University of Chicago Medicine maternity ward. In order to discern whether a mindset was already in place, new mothers had been asked if they agreed or disagreed with the statement: "How smart an infant will become depends mostly on his or her natural intelligence at birth."

While many of the mothers of all socioeconomic groups did not agree, the difference in those who did agree was highly significant, with new mothers from lower socioeconomic strata being far more likely to agree than their upper-strata counterparts.

The ultimate concern, of course, is that when parents believe that nothing can be done to positively influence a child's intellectual potential, the child is less likely to receive the necessary, additional support for intellectual development.

The question, however, is why there is a socioeconomic factor in this belief.

While the question is complex and an answer will take more than conjecture, my feelings are strong in this matter. I think that when people, either individually or in a group, have been told, over and over, in so many ways, that they are incapable of learning or doing, that belief is transmitted to the core. The concept of malleable intellectual growth never gets a chance. This is not to discount the many people who have been brought up with that weight and have surmounted it. But for many, the weight is so heavy and so energy sapping that it is an effective barrier to achievement.

While most of us have a little voice that says, "You're not going to be able to do it," those with the gift of persistence overcome it. But when that little voice is backed by a chorus of history saying, "Not you, you're just not smart enough, you'll never do it," reinforced by almost insurmountable societal restraints, the incentive to keep going can be drained completely.

That's what makes what happened next so inspiring.

A CHANGE OF VIEW

We met with the new mothers again for the development of our "newborn intervention." And we found a remarkable change.

Many of those mothers, the same mothers who had looked at their newborns as a book already written, now saw them as cute, loveable, malleable potentials. Potentials that they could be part of cultivating. While what we saw was anecdotal, not enough to claim statistical relevance, it was definitely enough to offer hope.

This hope sent me straight to the scientific literature. I wanted to see if there were studies demonstrating that changing parental "mindsets" altered parenting "culture." In other words, had anyone found that changing parents' views on fixed versus malleable abilities changed their parenting approaches, making them more consciously active in their children's endeavors?

The research had been done by Elizabeth Moorman (now Moorman Kim), PhD, currently a postdoctoral associate at the Nebraska Center for Research on Children, Youth, Families and Schools and Eva Pomerantz, PhD, a professor of psychology at the University of Illinois, who had studied this question with seventy-nine mothers of seven-and-a-half-year-old children.

Moorman Kim and Pomerantz hypothesized that a fixed mindset in parents would result in parenting practices that were less supportive of intellectual growth in children. In other words, parents who believed that intelligence could not be altered treated learning difficulties in their children as an indication of "fixed ability," with no chance of improvement. As a result, rather than offering constructive methods of learning, these mothers would encourage their children to "look good," including telling them how to solve problems rather than letting the children try to learn by themselves, thereby avoiding the stigma of failure. In these mothers there would also be observable frustration with their children.

Moorman Kim and Pomerantz theorized that engendering a growth mindset in parents would help them understand their children's ability as malleable rather than fixed. As a result, their approach to their children's struggling would be to view it as an opportunity to help their children learn how to learn, step-by-constructive-step, even when those steps were difficult to climb.

THE MOORMAN-POMERANTZ STUDY

Mothers were randomly assigned by the study to be a "growth mindset" or "fixed mindset" parent. All parents were then told that their child was to be given the Raven's matrices test for measuring intelligence.

Mothers assigned to the "fixed mindset" group were told: "The Raven's matrices tests your child's *innate, inborn* intelligence."

Mothers assigned the "growth mindset" group were told: "The Raven's matrices tests your child's intellectual *potential*."

All mothers were told they could help their children as little or as much as they liked during the test.

The test was rigged, way too difficult for any of the children. While the children struggled, the researchers watched to see how the mothers reacted.

Mothers who had been led to be "fixed mindset," that is, to

believe that their child had "innate" ability, were not constructive. But they *were* active, exerting observable control. They were much more likely to tell their children how to get the right answer rather than giving encouraging support while the child found it alone. Some mothers even took the pencil away from their child and completed the problem themselves. Fixed mindset mothers were also significantly more likely to use unconstructive parenting techniques, criticizing, for example, when children appeared helpless or frustrated, with responses that were compared to being hit when you are down.

On the other hand, an interesting finding was that "constructive" parenting practices did *not* turn out to be the flip side of unconstructive parenting practices. That is, providing a parent with a "growth mindset" frame did not automatically mean that constructive parenting techniques followed. It just meant somewhat fewer controlling, unconstructive techniques being used.

Why?

Because a parent's awareness that a child's intelligence is malleable doesn't mean that a parent has the techniques to make good on this knowledge. "Babies aren't born smart" does not automatically lead to "They're made smart by parents talking with them." You may be pointed in the right direction, but you still need to travel the road to the destination.

THE STORY OF THRECIA

Nothing Beats a Failure Except a Try.
                    —Threcia, Portia's Mother

Threcia often said to her children, Portia, Magellan, Pierre, Tony, Marcus, and Noelle, "Nothing beats a failure except a try." Without one book on brain development, or the data of a double-blind study on child rearing, Threcia was growth mind-set parenting personified. Tenacity, education, and expectations were at the heart of her child rearing.

Ms. T, as everyone in the neighborhood called her, had a seventh-grade education and spent her entire life working as a maid. A kind of spiritual precursor to Professors Carol Dweck, James Heckman, and Angela Duckworth, Ms. T pushed her children toward determined achievement in the face of seemingly unbeatable obstacles. If I believed in channeling, I'd say somehow this granddaughter of a slave had something to do with their work.

Ms. T was born in 1921, when the years of slavery were still close enough to touch. Struggling to rear six children in East St. Louis, she worked hard to shield them from the chaos of the outside world in a crowded apartment that had no telephone or television. When food was scarce, she used her rural "Tennessee upbringing" to stretch the family food budget by serving squirrel and raccoon that she bought from farmers and hunters.

But there was always something to read. On trips to Goodwill, specifically for that purpose, she would buy stacks of *Life* and *Look* and paperback books for a nickel each. In addition, to show that even the best parents embarrass their children, Ms. T would write letters to her children's teachers, filled with spelling and grammatical errors, prodding them to make sure her children were getting what they needed to fulfill their educational potential. Despite her own seventh-grade education, she was determined that nothing would hold her children back. Her own experience had not kept her from believing in herself and, by extension, her children. A rare, wonderful quality in any human. Lucky children.

Ms. T did *not* have a "fixed mindset!"

"My mother raised us with a strong sense of collective responsibility for the survival of the whole . . . that is, our family," says Threcia's daughter, Portia. "No one person's needs ever outweighed the needs of the whole. We had to stand together and love and support each other. My mother valued education and most of all hard work and transferred that to our entire family. We were raised with a strong sense of right and wrong. My mother . . . had high expectations for her children given her understanding of the world at that time."

It is not simply that Threcia had high expectations for her children; because of her, they had high expectations for themselves. She also made sure they had the critical tool to fulfill those expectations: education. Otherwise, what were all the *Life* magazines about? Her children knew, because she surreptitiously told them, that there was another world to know about, to become part of. "Here," she said, "read; that's where it's happening."

Threcia gave her children something more: the grit to persevere through all of life's challenges. "It's a good day to be Mrs. Jones's daughter" was sibling code for "I've had a really rough day." "But we all kept pushing forward with real tenacity," says Portia. "It was like we looked into the face of adversity and against all odds, kept plowing through."

Who is Threcia's daughter Portia? She's Portia Kennel, senior vice president of program innovation for the Ounce of Prevention Fund and executive director of the Educare Learning Network. Organizations that are prominent early childhood proponents, they do the double duty of working directly with parents as well as designing policies in early childhood. The first Educare early childhood center is now considered a national standard for high-quality learning. It was started by Portia Kennel.

To show that "growth mindset" can be transferred from parent to child, the first attempts in early childhood programming at Ounce of Prevention, called the Beethoven Project, were less than stellar. Portia could have given up or continued doing more of the same. But Portia, daughter of Threcia, knew what her end goal was, the enhancement of children's lives, and she knew Ounce of Prevention could do better. Scrapping the earlier program, she then designed the first incredibly effective Educare, which became a national model. If that isn't a growth mindset, I don't know what is.

I consider Ms. T's early death at sixty-five, after a long struggle with cancer, a tragedy. She never saw the complete fruits of her parenting labor. But if it's true that we live on in what good we have done, she's going to have a long, flowering, eternal future.

The child who was embarrassed by her mother's poor spelling and grammar is now an adult, pivotal in helping parents, often mothers, understand their role in their children's development, making them the strengtheners and supporters their children need. Portia is, indeed, solid proof of Ms. T's legacy.

A little aside: The name Portia did *not* come from *The Merchant of Venice.* Rather it's from the main character in the radio soap opera *Portia Faces Life,* which ran from 1940 to 1970, about a strong female attorney who fights for justice in the face of life's hardships. Our Portia was appropriately named.

Can we inspire a little Ms. T in every parent? Or do they already have it and just don't realize it? Those are hard questions. If you've grown up on a path carved out by your parents' lives, following right behind them, looking neither right nor left, nor up nor down, but just straight ahead, and your expectation is that this is the only path you were meant to be on, no matter if the sights are much lower than you might have wished for, how do you rear your own children to believe otherwise? How is a growth mindset imbued in parents who have never considered that there is an alternative to the fixed path of life?

When I broached this question with Portia, she laughed, pointing out that if Educare had existed when she was a child, her mother would have been the one walking through their doors.

SUCCESSFUL PARENTING

Author Wes Moore has said it best.

"We are products of our expectations," he writes. "Someone, at some point, put those expectations in our minds and we either live up to them, or live down to them. The only difference in my life was that there were people who were willing to hold on to my dreams long enough for me to grow and to mature and to find out that they, too, were my dreams."

What Wes Moore is saying is that along with the support of "growth mindset" when we're young, parents also need to be our rear guard, making sure that any backsliding goes only so far without someone to stop it. As someone once told me, "If you want your children not to be afraid to soar make sure they know that even if they fail, they can only fall so far before someone is there to catch them. That way, they'll try, try, try until it really works."

EDUCARE ALUMNI: MODERN THRECIAS

In 2012, Portia Kennel was asked to start an Educare Alumni Network by parents whose children had graduated from the program, some more than ten years earlier. The parents wanted to give back and to become catalysts for change in their own communities. Portia said the first meeting was exhilarating. Parents not only laid down the basic structure of the organization; they were filled with plans and ideas, ultimately laying the framework for a robust network that could have a broad positive influence on child care.

And then the full impact of what she had witnessed hit her: that Educare had not just been the support for children; it had been a life-changing, enlightening experience for parents, as well. To Portia it was inspirational.

Then a second truth hit her. After the first meeting with

alumni parents, Portia went back to her colleagues, telling them enthusiastically about what the parents had done and how much potential had been revealed in the parents. The responses were surprising. While some who listened to her were also exhilarated, others were blasé and relatively unresponsive to what she considered an extraordinary experience with these parents.

And that made Portia wonder. Had an unintended consequence of framing parents as the target for change somehow coated them with a negative aura? And had some of those who worked hard at encouraging "growth mindset" in parents developed, in themselves, a "fixed mindset" toward those parents? Was that the reason some were not able to see the incredible growth potential and actual growth the parents had exhibited?

"Don't get me wrong," Portia emphasized. "Our field is filled with the best of people, and what we do is incredibly important. I just wonder if, to a certain degree, we need a reframing of our thinking."

Which made me wonder, is there also a societal fixed mindset?

I had come to Portia to learn more about how to help transform parents' and caregivers' fixed mindset parenting into growth mindset parenting. I came away with answers, but with a lot more questions.

Is there, I wondered, a societal fixed mindset to entrenched societal problems? Have we assumed that because problems have existed for so long, they are immutable, unchangeable, and that there is no possible way they can be addressed? And is this a factor in the response to the need for policies to help improve them?

The science is indisputable. The essential years for developing the human brain are from birth through three years of age. That doesn't mean that the day you blow out four candles is the end of the brain's development, but they are critical years.

Science also tells us the key factors in that development. A child has to have adequate nutrition. A child has to have adequate language. Nature is kind; it demands and then supplies what is necessary to fulfill those demands. Almost every parent, without outside help, has the power to give a young child what is necessary for optimum development.

What keeps this from happening all of the time? We could parse together esoteric reasons, but ultimately it's probably because, while the awareness for food is second nature, the awareness of the need for rich language is fairly recent. The science is new; appreciation of it is new.

Nonetheless, even though we now know the need for an early language environment is imperative, the incentive for making sure it happens lags. Educational investment is almost always for children from preschool through twelfth grade. This is an important time, as well. But, as we've said, too often that expenditure is for solving existing problems. The roots of the problems in literacy, math, or even executive function, are clearly, science tells us, in the years between birth and three years of age. Solving these problems means a new concentrated effort in those early years, because their effect eventually becomes our country's achievement gap.

Professor James Heckman has written, "Traditional policy interventions fail to attack the root cause of the achievement gap. To equalize the playing field, governments need to invest in parents so parents can better invest in their children."

Ariel Kalil, professor of public policy at the University of

Chicago, has suggested that the limited support for parenting programs relative to early childhood education programs has another aspect, arising in part from a governmental view that families are not institutions to be regulated. Families, she has said, are viewed as private decision makers. But public policy has an important role to play, she goes on, in sharing the science of brain development and strategies for ensuring optimal child growth and development. Public policy initiatives of this sort should not be viewed as trying to change parental preferences, but rather as providing parents with the tools to help them achieve their own goals of raising happy, healthy, and productive adults.

HOW WILL CHANGE OCCUR?

To effect change, there has to be a conscious, universal effort to understand both the science and the eventual consequences for the child, for the adult that child becomes, and for the country in which that adult will function. Investment in early childhood has to have a new, strong thrust from a concerned population that understands the problem and the need for attention. This does not mean discarding current programs for older children. It means expanding them to the first day of life.

In other words, if we want to reap the greatest benefit from the money we invest in kindergarten through twelfth grade, we have to make sure that the children entering kindergarten are ready to learn, at their optimum level.

And it can happen. The first lady of Illinois, Diana Rauner, profoundly aware of the problems and the supporting science, is working to have a universal, supportive home visit for every baby born in Illinois. How's that for being proactive? And smart.

A SOCIETAL GROWTH MINDSET

There is no magic wand. Just believing in the intellectual malleability of all children does not mean we can make all children grow to their optimum potentials. There are many facets to the problems we see in achievement in our country, many things we, as a nation, must attend to, to help our population function at its best. But it's a good start.

Statistics tell us we have a problem: the gap in achievement among our children. Science shows us ways to solve the problem. This doesn't mean simply having one program replicated everywhere. But it does mean using science to accurately define the problem, then using science to help design programs that evolve, as ongoing review informs us, to correct it. In that way a serious, relentless problem can become part of this country's social history.

But only the population of this democracy can determine whether that will happen.

THE REQUISITES

We have to make the importance of the early language environment part of the American vernacular. Every parent, in fact, every person, should understand it. When parents want and need support, making sure that support is readily available should be second nature to us as a nation. And the programs that are designed should be scientifically sound, recognizing the importance of parents in the development of the young child.

We also have to recognize that when supportive programs are needed, and offered, it is not delineating differences in our population. Rather it is an affirmation that we, as a nation, diverse in every possible way, have a mutual commitment to making sure all our children can reach their optimum potentials, in intelligence, stability, and productivity, for their sake and our country's sake.

At one point we described birth as the luck of the draw. That luck does not extend only to the parents one was born to, but to the country into which one was born. We are a country of enormous potential, but only the positive involvement of our population can determine whether we meet that potential.

SCIENCE AS THE BASIS FOR REAL SOCIAL CHANGE

Science can be intimidating, an expertise that someone else has. But it shouldn't be. Because science simply refers to identifying a problem, breaking it down into understandable components, studying it, restudying it, and working back, step by arduous step, until you find its cause and, eventually, its solution.

According to Ron Haskins, co-director of the Brookings Center on Children and Families and Budgeting for National Priorities Project, a large majority of social service programs, costing billions and billions, have little or no effect. Many programs don't even collect data to determine whether they are working or not.

At TMW and at other programs working toward improving the outcomes for children, effectiveness is key. That is why science, not ideology, nor what we "believe," is core to our programs, both in identifying the problems and in designing, and honing, effective solutions. Our work does not stop in the face of questions or the need to reassess. Our ultimate goal is ensuring that all children have the chance to fulfill their potentials and that is what we, and our fellow organizations, are working to achieve.

Funding is, of course, a factor. Although we know that many of the problems embedded in the older child and the adult begin in the first three years, finding adequate resources to develop scientifically vetted interventions is often very difficult.

Jack Shonkoff and his colleagues are building a dynamic research and development platform, Frontiers of Innovation, a collaboration among researchers, practitioners, policy makers, investors, and experts in systems to design and test new ideas and learn from things that don't work, all aimed at breakthrough outcomes for young children facing adversity. Dr. Shonkoff is quoted as saying, "Transformational change requires entrepreneurial investments in science-based innovation, in addition to philanthropic support. . . . While improving quality and increasing access to best practices remains critically important, some segment of the field needs support for creative experimentation, implementation, evaluation, and sharing of knowledge about what doesn't work as well as what does. Venture-driven philanthropy is uniquely positioned to support this essential R&D dimension."

WHEN IT WORKS

My approach and that of the many others working in this field is absolutely growth mindset. Being director of research for child development, being a pediatric surgeon with all of the potential surprises of the operating room, has only reconfirmed what I know from simply dealing with the complexities of life: problems are solvable only with concerted, determined effort.

It is a mindset mirrored by the mothers in our initiative.

The most indelible memory I have of meeting with the mothers who were part of Thirty Million Words was how excited they were to be in a program designed to help build their children's brains. They knew it was a research project, that while we had strong, well-documented ideas on how to do it, that our initiative was designed to make sure they worked. Their enthusiasm helped build ours.

My admiration for these women only grew when I saw how much energy, physical and intellectual, it took to be part of

Thirty Million Words. Especially when it was clear how hard their lives were on this outer end of the socioeconomic spectrum. It's one thing to read about the struggles involved in being poor, but to live the stresses and the hardships can only be described as incredibly difficult. And even the word "difficult" is only skimming the surface. Which can only intensify respect for these mothers, who still have the motivation and determination to want to make life better for their children.

The ages of our mothers ranged from nineteen to forty-one, some with one child, others with two, three, or four. Some couch surfed from family member to family member, some lived in apartments in high-crime areas that made us hesitant about sending research assistants for home visits. There were, in fact, during the period of home visiting, episodes of violence, severe illness, and chaos experienced by the mothers and the children. But through it all, their resolve didn't waver. I have to thank these women; in fact, I *have* thanked these women, for tenacity I don't think I'd ever seen before.

While some of these mothers may have begun their parenting years with a fixed mindset about intelligence and learning, when they found out they could be the pivotal factor in their children's academic achievement, and the need for language, for positive reinforcement, and for stability was understood, they worked hard to make them part of their life's routines.

THE DUAL-GENERATION APPROACH

Developing a growth mindset does not, however, signify overnight success. There are tremendous obstacles related to poverty, income inequality, and opportunity gaps that impinge on both parents and children. A growth mindset approach is not "pull yourself up by your bootstraps" redressed. Rather, it is recognizing that there are untapped potentials in all of us and that with the right programs with the right support, there can be success.

One hindrance to the overall success of programs, both philanthropic and governmental, may be attributed to what Ellen Galinsky, president of the Families and Work Institute and author of *Mind in the Making*, calls the "twin streams." A pioneer whose research spans from early childhood to the adult workforce, Ellen Galinsky says that traditionally there has been a dichotomy between programs for parents and those for children. Agencies whose focus is children often work at the "expense" of the parents; workforce development/welfare reform programs are typically oriented toward adults, with much less thought for their children, often at the expense of the children. The result has been that one or the other is left unattended, with no help or support.

A dual-generation approach changes that by building simultaneous educational, economic, health, and security foundations geared to the stability and enhanced lives of parents and their children. Its basis is dependent on a distinctly growth mindset view of both parents and children.

When it was first used in the 1980s and 1990s, however, the dual-generational approach had less than stellar results. A glance at those results might lead to simply abandoning the idea. Further investigation has, however, provided important clues for how it might have been a striking success. These include initiating work-training programs rather than simple job placement, as well as offering programs that would help parents manage their dual roles as supportive parent and breadwinner.

The dual-generation approach is a component of the Community Action Project (CAP) of Tulsa directed by Steven Dow. One of the first dual-generation programs in the United States, its CareerAdvance enriches Tulsa's strong system of Early Head Start Centers and Head Start Centers with high-quality career-oriented training for parents in health care occupations including medical assistant, pharmacy technician, dental assistant, physical therapy assistant, and nurse. Parent education and training are done in partnership with Tulsa Community College and the Tulsa Technology Center.

With coordinated programming, CareerAdvance blends Early Head Start admissions for children with coaching support for parents who are entering the CareerAdvance program. Although Steve Dow and his team are doing admirable work, they still use scientific investigation to ascertain what works and what doesn't. As with every socially oriented program, all the answers are not yet in. But the one thing that seems certain is that the program is important, positive, and constructive for the children and the parents.

TMW'S DUAL-GENERATION
EXPERIENCE

Many TMW mothers have told us how much they hoped, after they completed TMW, that they could pursue their own education. It's possible that seeing their incredible power in helping their children flourish had reawakened their own dreams and, perhaps, changed their fixed mindset about their own potentials. That would be inspirational.

CHAPTER 7

SPREADING THE WORDS

THE NEXT STEP

You may never know what results come of your actions, but if you do nothing, there will be no results.

—attributed to Mahatma Gandhi

There is something about a country that cares enough to invest wisely in its children. And that something is stability, productiveness, and intelligent, constructive problem solving.

All people, all countries, have problems. The difference between people, and countries, is not whether they have problems, it's how they solve those problems. In a country where a great number of children cannot reach their highest potentials, the country cannot reach its highest potential, either. It is not that everyone has to think the same, but that ultimate conclusions are founded on solid rational thinking, not feeling, *thinking*, and for that you need brains well nurtured in early childhood followed by solid, excellent, accessible education.

HOW CAN WE MAKE THAT HAPPEN?

The earliest language environment is a key component to a child's eventual learning trajectory. In the United States, the achievement gap that divides those who are successful academically from those who perform poorly, or drop out, is large. "Large" is, in fact, a very understated term for this massive division.

While science has shown us the root cause of the achievement gap, something else is needed to ensure that effective solutions are put in place. All parents, in fact, all adults in this country, must begin to understand the problems, and the necessary solutions, so they become a part of our national dialogue and our national fabric.

Atul Gawande, in his insightful *New Yorker* article "Slow Ideas," explores the ways innovative thought becomes accepted. What makes an idea spread? What makes people accept a well-designed concept or disregard it? What makes us want to be part of encouraging it?

In the 1800s, two important advances in medicine were discovered: anesthesia and antisepsis. The first prevented excruciating pain and a wildly thrashing patient during surgery; the other prevented invisible germs from infecting the surgical wounds, infections that were so common surgeons believed that an oozing wound was part of the healing process. Both discoveries were unparalleled strides for the world of surgery and medicine. But only one took hold: anesthesia. Simple hand washing before surgery or changing one's surgical gown between cases just seemed like a waste of time.

The surgeon J.M.T. Finney recalled that when he was a trainee at Massachusetts General Hospital in the late 1800s, hand washing was still rare. While surgeons soaked their instruments in carbolic acid, they continued to operate in black frock coats stiffened with the blood and viscera of previous operations, "the badge of a busy practice." Why? What were the reasons differentiating the acceptance of these two concepts, anesthesia and antisepsis? As Gawande states, *visibility* and *immediacy*.

"One combatted a visible and immediate problem (pain); the other . . . an invisible problem (germs) whose effects wouldn't be manifest until well after the operation." This is, says Gawande, "the pattern of many important but stalled ideas."

HOW DOES THIS RELATE TO CHILDREN?

The achievement gap seen in students in kindergarten through twelfth grade is apparent to anyone who looks at the statistics. It's impossible to hide from them. It's also impossible to turn away from the effect on the adulthoods of these children.

Birth through three years is, on the other hand, a relatively invisible period. The achievement gap is already present at nine months of age but is apparent only under the microscope of statistical analysis. Without a concerted look, we might actually believe that the problems we see in older children start at the moment we observe them. Action has, as a result, traditionally been taken only after the problems become apparent. It took the prescience of Hart and Risley and the astute researchers who followed to show us that the problems existent in children of school age were simply observable manifestations of much earlier problems.

Knowing when a problem begins is not, however, the same as knowing what to do about it. Designing an appropriate solution demands finding out, first of all, why it occurs. Hart and Risley, even though they theorized that early language environments were the precipitating factor in poor school performance, had to back up their idea with sound statistical evidence.

But, as we saw, even discovering the cause of a problem does not necessarily lead to putting in place the requisite solutions to stop it. Just because physicians understood the relationship between infection and sepsis did not mean that washing hands or changing clothes was immediately incorporated into the general routine of surgery. It took time, even though doctors knew the science. Once the fact that invasive bacteria were the culprits in later, often fatal, infections became part of the *fabric* of surgical thought, however, surgery changed. Surgeons started washing thoroughly before going into the surgical theater and wearing sterile gloves and sterile "scrubs," things that everyone involved in an operation began to do. The results were indisputable and immediate, the outcomes improved beyond expectation. But it took time and, undoubtedly, lives.

The early language environment is critical for a young child's brain development. In order to ensure the optimal brain development of all our children, effective, well-designed support has to be readily available when there is a need. Before this can happen, however, general acceptance of the importance of the early language environment has to occur at a population level.

If this does not happen, it is, as Atul Gawande would character-ize it, a *slow idea* and, as such, an idea that will not lead soon to an effective solution.

AMERICA'S GREATEST UNDEVELOPED RESOURCE

The United States is replete with resources, including oil, gas, coal, copper, lead, molybdenum, phosphates, rare earth ele-ments, uranium, bauxite, gold, iron, mercury, nickel, potash, silver, tungsten, zinc, petroleum, natural gas, and timber. It has the world's largest coal reserves, accounting for 28 percent of the world's total. It is one of the world's largest national econo-mies.

And yet America's greatest resource needs attention: its chil-dren. To be an effective part of our increasingly globalized world, the United States is dependent on how well its citizens think, how thoroughly they analyze problems, how construc-tively they solve those problems. Today this country is depen-dent on us. Tomorrow there will be a new crop of citizens who will take our place in attempting to make this democracy pro-ductive, rational, and stable. We have a choice. We can help produce the highest quality of future citizens by working to en-sure the optimum development of our children. Or not.

AMERICA'S SECOND-GREATEST
RESOURCE

Parent talk, the quality and quantity of words in an early language environment, is an extraordinarily powerful and underutilized natural resource in this country and perhaps in most countries of the world.

Columbia University's National Center for Children in Poverty research showed that in 2013 approximately thirty-two million U.S. children lived in low-income families, sixteen million of whom lived below the poverty line. Although there are always exceptions, those are the children who will be least likely to go on educationally, whose academic and lifetime achievement prognosis is poor, and whose intellectual potential at birth is far greater than they, or we, will likely ever realize. The great preponderance of their parents, research has demonstrated, want their children to achieve educationally, but the incalculable stresses of poverty, both personal and societal, coupled with a lack of appropriate support, often keeps it from happening.

It does not have to be that way. While all the answers are not in, right now, in this country, we have what we need to begin improving outcomes for our children and for this country's future. In fact, once we begin, with carefully designed, carefully monitored programs, increasingly definitive answers will become apparent. What it needs is appropriate investment. And it's a wise investment. While there may be reasonable debate about its precise value, Nobel laureate James Heckman has found that every dollar invested in quality early childhood education for disadvantaged children delivers annual economic gains of 7 percent to 10 percent through increased school achievement, healthy behavior, and adult productivity.

But this book is simply words on paper without universal involvement. While our understanding the problem is a first step, a long-term solution needs the attention of everyone. Only if we work together can we help ensure that programs, well designed and scientifically honed, are put into place to enhance the outcomes of all our children.

Who are "we"? We are the individuals who understand the problem and are the guardians of this important goal, actively supporting it. We are the organizations with initiatives that provide language programs for families and children, refining protocols as needed to make sure of their success. We are the public and private partnerships, small and large, offering support systems for families who want and need it. We are the groups providing information so that all parents understand the importance of the language environment of a child in the first three years.

Above all, we are the ones who do not simply believe in this, but rely on science to define the problem and help us design effective solutions. If there is passion, it is for ensuring that every child has the opportunity to achieve to his or her highest potential. We are not daunted when programs are not perfect; it just propels us to improve them to optimum success. Our ultimate goal is the enhanced lives of the children.

How can we achieve a universal understanding of the power of parent talk? While I began thinking about this when we started our program in 2007, my thinking took a fast-forward step in the fall of 2013.

In 2013, the White House Office of Science Technology Policy asked my team and me to help organize the conference "Bridging the Thirty-Million-Word Gap." It was to be done in partnership with the U.S. Department of Health and Human Services, the White House Office of Science and Technology Policy, the White House Office of Social Innovation and Civic Participation, and the U.S. Department of Education. Its purpose was to assemble researchers, practitioners, funders, policy makers, and thought leaders from around the country to discuss caregiver interventions and other solutions designed to help solve the problem of the U.S. achievement gap.

The conference was partly the result of interest generated by *Nudge,* the book by Professors Richard Thaler and Cass Sunstein. Growing out of the world of behavioral economics, the Nudge theory suggested that small tweaks or societal "nudges" can encourage positive population behaviors. Nudge theory, they showed, could be applied to everything from smoking during pregnancy to insulating attics to giving to charities. In his *New York Times* article, "Public Policies, Made to Fit People," Richard Thaler describes using a "behavioral nudge" as a method for closing the thirty million word gap. Our program was highlighted in the article, as was Providence Talks, a citywide home-visiting initiative that had been awarded the grand prize by the Bloomberg Mayors Challenge.

Ironically, while the conference was to have been a joint program with government counterparts, a last-minute government sequestration stripped the meeting of any federal partners. Nonetheless, the conference turned out fine. There was, in fact, a strong sense of shared purpose at the meeting. Many of the incredible social scientists mentioned in this book were there. To have so many dedicated researchers, practitioners, policy makers, and funders in the same room actively focusing on the same important issue, how to bridge the discrepancy in language acquisition/exposure, or the "word gap," and its devastating consequences, was inspiring.

The Nudge theory is, in fact, very interesting. Richard Thaler and Cass Sunstein's concept gives great encouragement that small behavioral nudges can impact language behavior in parents as a preliminary step in solving this problem. Making the change self-sustaining on a population level, I thought, might require an additional, more dynamic, thrust. Recognizing this started me on the path to defining my ultimate goals for TMW, including what a population-level shift might look like.

I had never viewed the initial iteration of the TMW home-visiting program as the program's final goal. However, for the power of parent talk to become embedded in the social fabric, I began to realize that its importance had to become part of the national conversation, including obstetrical clinics, maternity wards, physicians' offices, the early care curricula and, especially, from parent to parent to parent. That vision was transmitted in our conference paper "Bridging the Early Language Gap: A Plan for Scaling Up."

SPREADING THE MESSAGE

If language and the power of parent talk are to be universally understood as the essential nutrition for the developing brain, they have to become an integral part of public thought and part of the culture of early child care in a surround sound that every parent hears: "Talk with your baby, talk nicely with your baby, elicit from your baby."

It's important to stress that we are not talking about changing idiomatic speech or cultural linguistics. Early language interventions do not require people to change the words they use, nor do they ever denigrate customary language practices. Rather they focus on enriching those parent-child interactions that promote school readiness, including conversational turns and responsive language, encouraging parents to use the language, speech patterns, and stories most natural to them. A successful, population-inclusive intervention strategy will include videos, pictures, songs, and narratives from across our beautifully diverse population with its myriad of cultural, ethnic, and racial backgrounds.

A KEY PUBLIC HEALTH INDICATOR

The United States pays attention to public health indicators including vaccination and prematurity rates. If a child's early language environment is a critical catalyst for brain growth, then the early language environments of children from birth through three or five years of age have to be considered a barometer for the health of this nation and should be followed, as well. The use of specially designed technology, similar to the LENA, could support it as a viable public health approach.

One reason this is not currently done is because it's much

easier to follow children later in life, when they are in a school setting. But nearly twelve million children under five years of age are in some form of center-based child care. These settings would be appropriate for monitoring the language environment of young children, including assessing variables for long-term learning. Stay-at-home caregivers could also be given the opportunity to measure the language environment of their children on a voluntary basis.

The early learning community recognizes the importance of measuring and promoting the quality of the early language environment. But there are, says Ann Hanson, director of advancing quality at the Ounce of Prevention Fund, significant challenges to making this happen. "Currently, we monitor and measure many important indicators of quality in early learning programs, from classroom structures and caregiver qualifications to teacher-child interactions. But the real opportunity is to focus on the things that matter most. If science tells us that the early language environment is foundational to development, [we also must know] what type of tools and supports will give educators the appropriate, timely, useful data and strategies to improve it."

Ann Hanson notes that another drawback has been that while widely used assessments for the quality of learning environments, including language, exist, their effects are limited because typically they are conducted only annually. Defining the early language environment as a key public health indicator could lead to timely data useful in creating guidelines for the development and improvement of early language programs. These validated standards for language environments could also be incorporated into state early learning standards, providing parameters for quality assessment of child care facilities and guidelines for improvement.

Professor Gail Joseph, associate professor at the University of Washington and director of the Childcare Quality and Early Learning Center for Research and Professional Development, is addressing this issue by studying the language environments of child care settings. While it is early, she and her colleagues, using the LENA, are finding a positive relationship between child care provider language with children, both the number of words spoken and the length of back-and-forth conversations, and key child outcomes. She hopes to identify the parameters of an optimal language environment that can be used for assessing child care quality. These validated standards could also be incorporated into state early learning standards, providing parameters for quality assessment of child care facilities and guidelines for improvement.

Defining optimum language environments for young children could also help in the design of training programs for early childhood providers. The parameters could be included in the standards for the Child Development Associate Credential, the entry-level qualification for early childhood education, and be integrated into other programs geared to children in early learning settings. One important result would be to help reassure the parents of the millions of children in child care that their children were being cared for in a rich early language environment. They could also offer guidance to families receiving home-visiting services. Importantly, as a public health approach, they would be available to everyone, in every community, across all socioeconomic strata.

THE HEALTH CARE SYSTEM

The health care system, which serves the medical needs of nearly all our children, is a logical platform for teaching parents the importance of the early language environment. Ideally, that's exactly what it would do. But the ideal does not always fit comfortably into reality.

According to Dr. Perri Klass, pediatrician, author, and medical director of the national Reach Out and Read initiative, primary care pediatricians and nurses understand the critical importance of guiding parents in their role of helping their children develop cognitively. They give advice, "anticipatory guidance" is the term used, about the ways children are expected to change as they grow and develop, and how parents can foster healthy and safe growth and development. But these conversations take time, and limited time encroaches on even the best of intentions in our fee-for-service world of medicine. In many practices and clinics, pediatricians are under pressure to show volume of patients seen, and areas that do not seem of imminent concern, including the developmental "anticipatory guidance," which includes helping parents understand the role of the language environment on ultimate development, take on an "only if there's time" status.

"We all feel pressed for time," states Dr. Klass. "We have so many things to check and we all lose sleep over missing the pathologic or rare diagnosis . . . the one leukemia patient. But we also know that anticipatory guidance for behavioral and developmental issues are critical for many children. We need to find a way to do both in our limited time."

HOPE

The first official White House meeting to "Bridge the Word Gap," organized by Maya Shankar and her colleagues at the White House Office of Science Technology Policy, was held in partnership with Too Small to Fail and the Urban Institute in October 2014. With so many organizations confirming their commitment to "closing the word gap," and the administration's announcing its support on this critical issue, the energy was wonderful. A grant, funded by the U.S. Department of Health and Human Services for a "Bridging the Word Gap Research Network" was awarded, with historical justice, to the University of Kansas Juniper Gardens Project, including Professor Dale Walker, who had done the Hart and Risley third-grade follow-up study. She with fellow researchers Professors Judith Carta and Charlie Greenwood, the scientific descendants of Hart and Risley, are continuing this pivotal research in their communities.

The programs now working to solve the problem of poor academic achievement in our children are doing an incredible job. The ones listed below are prime examples. They, and many others, are described in more detail in our appendix.

Educare
Mind in the Making
Providence Talks
Reach Out and Read
Talk With Me Baby
Too Small to Fail
Vroom

Programs such as these focus on empowering parents as the pivotal factors in the optimal development of their children. They lay a well-honed groundwork for broader-based, federally supported programs that could be integrated into the national agenda to help ensure school readiness and long-term academic and personal achievement in all our children.

THE THIRTY MILLION WORDS INITIATIVE

TMW's ultimate goals are a universal appreciation for the need to improve the early language environments of children and a national momentum for supporting the programs that will make it happen. While we are passionate in our determination to ensure that every child has the opportunity to fulfill his or her potentials, rational science is what guides us.

TMW's research is geared toward developing evidence-based curricula that could be incorporated into existing settings, including newborn nurseries, pediatricians' offices, home-visiting programs, child care programs, and community organizations.

While TMW's design would allow it to be adapted to site-specific needs, the foundational principle would remain the same: Children aren't *born* smart; they're *made* smart by parents' and caregivers' talk. The Three Ts: Tune In, Talk More, and Take Turns, would remain the core method for enriching a child's early language environment.

An important adjunct to this would be establishing parent language, and its importance, in the universal vernacular, so that when a pediatrician, the maternity nurse, or an early childhood teacher talks about using the Three Ts, parents immediately understand. Professionals, including early learning and child care workers, could be taught about the Three Ts in training or online, helping them understand the critical importance of their language with the children in their care. The interaction of health care and educational professionals, child care workers, and parents would help to establish an interwoven community that can ultimately be a cultural foundation of intellectual growth for children.

Technology could help in many ways as well, acting, as it often does, as a galvanizing force in a population-wide understanding of a program. A computer-based platform for our curriculum has other advantages as well, including embedded technology that could help measure the impact of various strategies so that techniques could be evaluated and, when necessary, honed. This would be done in a way that is anonymous but helps perfect the programs. Envision a data-driven Thirty Million Words initiative, supported by an interactive web design, a Khan Academy of sorts, that offered free, accessible, evidence-based early language programs for parents of babies and young children.

WHEN WILL LASTING CHANGE OCCUR?

All the pediatricians, all the health care workers, all the teachers in the world knowing the importance of language in a child's first three years means nothing if the parents don't know. A child's early language environment is dependent on the parent or the primary caregiver. Without them, the necessary growth will simply not occur. When I began Thirty Million Words, I would look at the babies' heads and imagine the rapid firing of developing neurons just at that moment. Now I look at the adults who care for them and think, "You are more powerful than you ever imagined and I hope you know it."

When our first pilot study for the TMW home-visiting curriculum was over, we brought the mothers together to get their feedback about what worked, what didn't work, what could be done differently. Our parents are active partners in the development of TMW and their input is essential to shaping the next iteration of our home studies.

We got a wealth of information. These women who had never met before, home visiting is conducted house by house, were like an established committee, bonding as if they had known one another forever. It was apparent that they recognized their importance to the study and especially the need for honest evaluation to make sure the program worked well. There was a kind of social bonding as they discussed things with one another in a back-and-forth refining of decisions. When they presented their ideas to us, it was with the full intent of helping to shape the procedures that would be followed in the next round of TMW, a study of which they felt an integral part.

Some of what they discussed described what they had learned and how they had incorporated that knowledge into their child rearing, including talking with their children even when they were almost too tired to talk. Even mothers who, early on, had low LENA scores, were talking like seasoned, successful veterans of the program. It's amazing what positive social reinforcement makes happen. The dialogue was inspiring to them and to us. While they gave us the definitive feedback we needed for enhancing the curriculum, they didn't stop there, telling us how to publicize and, even more impressively, *why* to publicize. To show how prescient these women were, remember that this was several years before the "close the word gap" momentum. They were definitely ahead of their time in being creative, integral parts of the movement and, even more, in recognizing its need.

But I recognized something they didn't. That, while they talked about using billboards and WIC offices to publicize TMW, *they* were really the most powerful method for spreading the message. And I was right. Later we discovered that these mothers had not only shared TMW's information with their coworkers and church members, some had even taught their siblings with young children the Three Ts so they could use it as well.

Professors Noshir Contractor and Leslie DeChurch in their article "Integrating Social Networks and Human Social Motives to Achieve Social Influence at Scale" describe what it takes to move "scientific discoveries to public good." Their research goal was to develop a framework for allowing important scientifically based ideas to take root within a community. They wrote that in order for innovations, even those strongly rooted in science, to spread into general acceptance, they must go from "accepted truths on the part of a few scientists to commonplace beliefs and norms in the minds of the many." Contractor and DeChurch also traced the influence of "opinion leaders" on the attitudes and behaviors of a community, in speeding behavior change, and on the acceptance of innovative thought. Opinion leaders were defined as the people and groups "whose buy-in . . . seeds cascades of attitude change and new norms within the community." In other words, those who make Atul Gawande's "slow ideas" transform into "fast" ones.

The importance of these mothers in "spreading the words," therefore, cannot be overestimated.

SPREADING THE WORDS

Spread the Words is also a TMW component. It views every parent as an important opinion leader, the critical component in changing attitudes, in the acceptance of well-founded innovation, and an essential part of the solution. While Spread the Words has become a much more intentional and developed part of our program, it was when TMW was in its infancy that James made me understand how really effective one person can be.

JAMES

"Why do I tell my friends?" said James. "I tell them because I want their kids to have the same advantage that my kid has. I wouldn't want Marcus to be the only kid that knows these things or has the upper advantage."

That's what James said to the team at the end of his time with TMW when he was discussing how he disseminated the information about TMW to others. Probably some of the most altruistic, socially minded words I'd ever heard. No, James didn't want his son to be "better than the rest." No, he didn't want his son to have more than everyone else. He wanted everyone's child to have what he wanted for his own.

A tall man in his early twenties, James has a high school diploma, a love of music, a job as a stocker at Walmart, and a new, entrenched knowledge about how to make his only child's brain the best it can be. Not content with casually "spreading the words," James routinely Skyped with friends in Atlanta and Indianapolis, talked to his son's day care provider, and even recruited his brother to take part in the program. If Spread the Words began as a part of TMW, it was now an intrinsic part of James. While his message wasn't always TMW's message verbatim, it was always clear and always constructive.

I met James and his son Marcus in my ENT clinic, where they came regularly for Marcus's ear infections and chronic breathing issues. No question about father-son love; Marcus, about thirteen months old when I first met them, was very, very attached to his daddy. It's rare, but I actually remember the first time I met them. It wasn't just because it's generally moms who bring their children in, regardless of their socioeconomic status, or because little Marcus was always pristinely dressed, in miniature Nikes matching his father's even before he could walk by himself. It was the way James so absolutely adored Marcus; his pride in being his father was palpable.

"He's always smiling, playing, laughing. He yells a lot; he likes to just be the life of everything. He's my life. He makes me smile every day when I wake up," James said. "I never know when he's going to actually say his first word, or if he's going to wake up and do a math problem or something. It's just amazing."

He takes a breath.

"I honestly wasn't ready to be a dad, but as soon as he came, my whole life changed and I had to immediately grow up. From day one, February twelfth, I did everything that I could pretty much do to make him better than what I was as a child, and give him a head start and advantages that I didn't have when I was a child."

Rarely do families who go through TMW come from my clinical practice. But there was something about James, his relationship to Marcus and his philosophy of life, that made me ask him, during one visit, if he wanted to learn more about how he might help build Marcus's brain. Later, when I asked him why he'd accepted my offer, he said, "I guess because it would help me develop me develop him." A perfect answer.

Even though James had to maneuver TMW sessions around his Walmart shifts, he somehow made it work. And James absorbed TMW like a sponge.

"Thirty Million Words taught me that tuning in with the

kid, with my son, Marcus, when he's on the floor, probably interacting with his toy piano, that I should cut off all electronics, my cell phone, my computer, my TV, and actually get down and tune in with him. I'll show him B flat and C sharp and . . . different keys. When he plays on his drum, I sit down and I play on the drum with him. Tuning in pretty much has taught me how to get into what he's into and then teach me a lot about him while I'm learning as well. It is very cool that I can build my baby's brain. If he's babbling, sometimes actually says something, or kind of repeats when I'm reading to him, or actually pays attention when we're playing on his piano or looks at something I'm describing, touches it then looks back at me like 'is this what you're talking about?' It's just . . . well . . . wonderful."

None of what he said, because I knew James, surprised me. What did surprise me was how this totally cool guy started Spreading the Words, actively and intentionally, almost from the beginning.

Aaron was James's first recruit.

"I told my younger brother Aaron about TMW. When I first told him that when I was at home with Marcus, I turned off every bit of electronic equipment, including cell phones, I could tell he didn't really believe me. Then I showed him, getting down on the floor and tuning in, the way we're supposed to, with Marcus. Aaron's face changed completely. Then, as I started tuning in to what Marcus was doing, it was pretty much like, the proof was in the pudding. Aaron was hooked. From then on, Aaron's just been coming to the TMW sessions with me and now he's doing the things he's learned with his son."

"I have a lot of friends that have young kids and I share what

I've learned at TMW with them . . . all the Tune In, Take Turns, and Talk More concepts. I've taught them what I learned and now they're doing it with their kids, like talking about shapes and doing math and things. My friend in Georgia, Mora, we Skype. I've taught her the three Ts, you know, Talk More, Take Turns, and Tune In. Now she uses them on her little boy. Jeanie's my friend in Indianapolis. Skype with her, too. And taught her the same things. She uses those on her daughter, and she gets really descriptive with her and uses lots of words. The truth is, once they knew about the program they were kind of thirsty to know more about it, maybe sorry they don't have the same thing. So here's what I did. Every time I learned something I'd get on Skype to teach them."

And James didn't limit Spreading the Words to his friends.

"I also told Marcus's day care teacher about TMW. She knew a bit about it, but not about Tuning In or the fact that watching TV really doesn't teach words in a way that they stick. When I learned new things, I always brought them to her and she would start using them at the day care, like reading to the kids before their nap and when they ate. When she took them on nature walks if one of the kids picked up a leaf or something, she'd describe the leaf and talk about where the leaf came from and different things that would interest the kids."

"I think spreading the words about TMW and how powerful parent talk is, is important, because when I tell one friend and then that friend goes and tells another friend and then that friend tells a whole group of people, it's kind of like the domino effect. And then pretty soon we'll have a world full of smart babies running around."

James's love for his son was always there, but as he went through the program, his confidence in rearing Marcus grew and with it a feeling of empowerment and confidence in his child rearing. It was a confidence, I suspect, that was transmitted to others.

James illustrates what happens when parents understand their power in their children's future. He also exemplifies what happens when resources are available to parents who might want and need them. James is more than a good parent, he represents the importance of our goals, which includes making parents an integral part of the solution.

THE MOST IMPORTANT WORDS

We are a nation of incredible resources, but we are a nation with a serious problem, a problem that has both humane and pragmatic implications. Too many of our children face a future of unreachable potentials, affecting them, our country, and the world in which they will live.

We know the problem; we know the solutions; we know what we should begin to do.

Almost all parents in this country can give their children the necessary language environment to build their children's brains to optimum potential.

There is no child in this country who should be without the necessary language environment to build his or her brain to optimum potential.

If all parents, everywhere, understood that a word spoken to a young child is not simply a word but a building block for that child's brain, nurturing a stable, empathetic, intelligent adult, and had the support to make it happen, what a different world this would be.

A country that wants to reach its potential must make sure its population can reach its potential. Support for children, parents, and communities, including stable, safe housing, employment opportunities, adequate health care and, of course, well-designed early childhood programs, are important components to that end.

For the sake of our children; for the sake of our country; for the sake of our world, we must make it happen.

Together we can make it happen.

STEPPING OFF THE SHORELINE

T he waves rose six feet into the horizon over Lake Michigan. Our three children were playing in the sand, watched over by my husband, their father, Don Liu. As he stood at the shoreline he suddenly noticed in the turbulent, chaotic distance two young boys struggling in the raging waters. He started running into the lake as our younger daughter cried out, "Dad, don't go!"

They were the last words she ever said to her father. The two boys got back alive. My husband, always fearless when it came to helping others, died, overwhelmed by the torrential pounding of the waves and the gripping undertow. He was my best friend, my strongest support, my true love.

For Don, standing on the shore watching two children struggle, there was no debating, no hesitation. He was a pediatric surgeon, a leader in his field, and his devotion to his patients was unquestioned. A child needed help; the child got help. It was not simply a maxim; it was his way of life. He would never have considered standing on the shoreline while two children struggled, even if he knew taking action would cost him his life.

In our country, we have too many children who are struggling against the odds of achievement, children who came out of the womb not knowing what they should have demanded, in writing, to make sure their lives would match their potentials. They are floundering. We cannot stand on the shoreline.

Afterward, Don was lauded a hero. And that's what we all have to become.

Dedicated

To

Donald Liu, MD

1962–2012

APPENDIX

EARLY CHILDHOOD ORGANIZATIONS AND RESOURCES

WHAT'S HAPPENING TO GIVE US HOPE

Extraordinary initiatives are currently active or being developed throughout our country to help solve the problem of achievement in our children.

Too Small To Fail

The "Talking Is Teaching" campaign of Too Small to Fail incorporates the tagline "talk, read, sing." A joint venture of the not-for-profit Next Generation and the Clinton Foundation, Too Small to Fail includes among its partners major television producers including Univision, Text4baby, Sesame Workshop, the American Academy of Pediatrics, as well as others. Too Small to Fail launched a text-to-parents program in partnership with Text4baby and the Sesame Workshop to distribute research-based tips to new parents about the importance of talking, reading, and singing with their newborn children. The service is estimated to reach 820,000 parents nationwide. In a creative use of television, their message has been incorporated into the popular series *Orange Is the New Black* as well as in *Sesame Street,* where science-based talk tips are given to parents.

Talk With Me Baby

Georgia's statewide public health and education initiative, Talk With Me Baby (TWMB), aims to transform parents and caregivers into "conversational partners" with their infants in order to nourish the critical brain development required for higher learning. TWMB is integrating "language nutrition" coaching as an important component for professionals in child care, including nurses and WIC nutritionists already working with parents and babies.

This innovative effort is the result of the collaboration between the Georgia Department of Public Health, which has recognized language acquisition as a public health issue, the Georgia Department of Education, the Atlanta Speech School, Emory University's School of Nursing and Department of Pediatrics, the Marcus Autism Center at Children's Healthcare of Atlanta, and Get Georgia Reading, Georgia's Campaign for Grade-Level Reading.

Reach Out And Read

The National Reach Out and Read program, a not-for-profit founded in 1989, trains and supports medical providers, including pediatricians, family physicians and nurses, to advise parents at regular checkups about the importance of reading aloud to children, including providing families with age-appropriate children's books. Reach Out and Read incorporates five thousand clinics, health centers, and practices in all fifty states, and every year distributes 6.5 million books to more than four million children. Their data demonstrate a significant impact on child outcomes; children in their program scored three to six months ahead on vocabulary tests compared to those who were not in Reach Out and Read in their preschool years.

Educare

The purpose of Educare, created by the Ounce of Prevention Fund, is to provide a program, a place, a partnership, and a platform for early childhood education. Children deemed at risk for school failure are offered full-day, year-round instruction from birth to five years of age. The results have been very positive. Children who participated in Educare for two or more years performed at the same level as other children beginning kindergarten, as defined by established national averages.

Educare's foundation is science. It relies heavily on documented research data for both implementation of teaching practices and methods of evaluating progress. Its program includes well-trained early childhood educators whose goal is to help parents produce a healthy parent-child relationship as a requisite for optimizing a child's development. Parental involvement with Educare begins prenatally, continuing throughout the first five years of a child's life. The intense program includes strategies to enhance the child's learning and social-emotional development. Once the child is in school, Educare continues its involvement via social workers and early interventionists to help parents get access to the community and on-site resources they need to be successful. Educare reports that its parents are more likely to participate in school activities and to discuss their child's learning with teachers.

Mind In The Making

Led by Ellen Galinsky, president of the Families and Work Institute, Mind in the Making shares the science of children's learning with the general public, families, and professionals, and has been instrumental in helping parents understand the importance of self-regulation as an executive function. It is based on what Galinsky calls the "seven essential life skills every child needs," including self-control, perspective, communicating, making connections, critical thinking, taking on challenges, and self-directed, engaged learning. It empowers adults with strategies and skills to help develop a child's executive function and cognitive skills. Projects include Seven Essential Skills Learning Modules, which are being implemented in fifteen communities and states; a DVD collection of forty-two videos showing important experiments in child development research; Prescriptions for Learning, which are downloadable tip sheets for families and professionals on turning everyday behavioral challenges into opportunities for promoting life skills including executive function; and a partnership with First Book, with a library of more than one hundred children's books and accompanying tips to promote life skills.

Vroom

Vroom, funded and created by the Bezos Family Foundation, starts with the premise "Every parent has what it takes to be a brain builder." Materials include tools for community-based organizations and agencies that demystify the science of brain development; consumer products that include a brain-building prompt on the packages of common purchases; and a free mobile app. Parents who download the app are asked to put in their child's age, enabling the app to offer advice specific to a child's need. Information includes a "Daily Vroom," designed to make ordinary activities such as bath time or meal time, times for enhancing brain development and executive function. Vroom activities, above all, are geared to promoting positive interaction between the parent and child.

Providence Talks

Providence Talks is a home-visiting early intervention program that uses the LENA technology and biweekly coaching to help parents enrich their children's early language environment. The program was the grand prizewinner in the Bloomberg Philanthropies' 2012 Mayors Challenge. Providence Talks is partnering with Brown University to assess the impact of its citywide program.

Boston Basics Campaign

A Massachusetts program, the Boston Basics Campaign was convened by the Black Philanthropy Fund in association with the Mayor's Education Cabinet and the Achievement Gap Initiative (AGI) at Harvard University. The Campaign is organized around five propositions (the Boston Basics) for early childhood parenting and care giving distilled from research literature by the AGI with support from a national advisory council of leading researchers and input from Boston's early learning community. A growing coalition of partner agencies, including WBGH Public Broadcasting, the Dudley Street Neighborhood Initiative, and a host of early education and parenting services providers are collaborating to make the Boston Basics central to early childhood care giving in Boston.

Get Involved!

Alliance for Early Success
    http://earlysuccess.org/

Campaign for Grade Level Reading
    http://www.Gradelevelreading.net#sthash.PgI8D6HI.dpuf

Council for a Strong America
  www.councilforastrongamerica.org

First Five Years Fund
  http://ffyf.org/

First Focus
  www.firstfocus.org

National Black Child Development Institute
  www.nbcdi.org

National Council of La Raza
  http://www.nclr.org/index.php/issues_and_programs/education
  /ece/

Pew Center on the States Home Visiting Projects
  www.pewstates.org/projects/home-visiting-campaign
  -328065#sthash.PgI8D6HI.dpuf

The Ounce of Prevention Fund
  http://www.theounce.org/

Save the Children
  http://www.savethechildren.org
  /site/c.8rKLIXMGIpI4E/b.6115947/k.B143/Official_USA_Site.htm

Voices for Children
  http://www.voicesforchildren.org/

Parent Development And Leadership Programs

Abriendo Puertas
  http://ap-od.org/

AVANCE
  http://www.avance.org/

Centering Pregnancy and Centering Parenting
    https://www.centeringhealthcare.org/pages/centering-model
    /parenting-overview.php

Family Wellness
    www.familywellness.com/skills.php

Legacy for Children
    http://www.cdc.gov/ncbddd/childdevelopment/legacy.html

MALDEF (Mexican American Legal Defense Fund) Parent School
    Partnership Program
    http://www.maldef.org/leadership/programs/psp/index.html

Parent Empowerment Project
    http://www.nbcdi.org/what-we-do/parent-empowerment-project

Parent Leadership Training Institute
    http://www.cga.ct.gov/coc/plti_overview.htm

The Positive Parenting Program (Triple P)
    http://www.triplep.net/glo-en/home/

Strengthening Families / Parent Cafés
    http://www.bestrongfamilies.net/build-protective-factors
    /parent-cafcs/parent-cafe-model/

Strengthening Families Program
    http://www.strengtheningfamiliesprogram.org/

Model Early Education Programs

CAP Tulsa: Early Childhood Education
    https://captulsa.org/families/early-childhood-education/

David C. Abbott Early Learning Center
    http://www.marlboro.k12.nj.us/AbbottEarlyLearningCenter
    .cfm?subpage=11

Educare
>   http://www.theounce.org/educare

First 5 California
>   http://www.first5california.com/

LAUP
>   http://laup.net/

Providence Talks
>   http://www.providencetalks.org/

Project EAGLE, University of Kansas Medical Center
>   http://www.projecteagle.org/

Southwest Human Development
>   http://www.swhd.org/

Talk, Read, Play (Kansas City, Kansas, and Kansas City, Missouri)
>   https://www.thefamilyconservancy.org/parenting-resources
>   /talk-read-play

Talk with Your Baby
>   http://talkwithyourbaby.org/

Touch, Talk, Read, Play
>   http://www.urbanchildinstitute.org/key-initiatives/touch-talk
>   -read-play

United Way Center for Excellence in Early Education
>   http://www.unitedwaycfe.org/

Home-Visiting Programs

Any Baby Can
>   http://www.anybabycan.org/

Bright Beginnings (Colorado)
    http://brightbeginningsusa.org/denver-talks-back/

Early Head Start Home Visiting
    http://homvee.acf.hhs.gov/Implementation/3/Early-Head-Start
    -Home-Visiting-EHS-HV-/8

Healthy Families America
    http://www.healthyfamiliesamerica.org/home/index.shtml

Maternal, Infant, and Early Childhood Home Visiting
    http://mchb.hrsa.gov/programs/homevisiting/

Nurse-Family Partnership
    http://www.nursefamilypartnership.org/public-policy/Home
    -visiting-support-letters

Parents as Teachers
    http://www.parentsasteachers.org/

Resources For Parents And Educators

American Academy of Pediatrics
    http://www.aap.org

American Library Association
    http://www.ala.org

Ascend: Developing Kids to Their Full Potential
    http://ascendkids.com/

Aspen Institute Ascend Network
    http://ascend.aspeninstitute.org/network

Baby Talk
    http://www.usfsp.edu/fsc/baby-talk/

Brazelton Touchpoints
    http://www.brazeltontouchpoints.org/

Child Care Aware of America
    http://childcareaware.org/

Early Childhood Technical Assistance Center
    http://ectacenter.org/

Get Ready to Read (National Center for Learning Disabilities)
    http://www.getreadytoread.org/early-learning-childhood
    -basics/early-childhood/understanding-language-development
    -in-preschoolers

The Hanen Program for Parents of Children with Language Delays:
    It Takes Two to Talk
    http://www.hanen.org/Programs/For-Parents/It-Takes-Two-to
    -Talk.aspx

Illinois Early Learning Project
    http://illinoisearlylearning.org/

The Incredible Years
    http://incredibleyears.com/

Learn The Signs. Act Early. (CDC)
    http://www.cdc.gov/ncbddd/actearly/

National Association for the Education of Young Children
    www.naeyc.org

National Association for Family Child Care
    http://www.nafcc.org

National Center for Families Learning
    http://www.familieslearning.org/

National Center on Parent, Family, and Community Engagement
    http://eclkc.ohs.acf.hhs.gov/hslc/tta-system/family

National Head Start Association
    http://www.nhsa.org/

PBS Parents
http://www.pbs.org/parents/

Baby & Toddler
http://www.pbs.org/parents/child-development/baby-and
-toddler/

PNC Grow Up Great
https://www.pnc.com/grow-up-great

Reach Out and Read
http://www.reachoutandread.org/

http://www.reachoutandread.org/why-we-work/our-initiatives/

Read-Talk-Play
http://www.readtalkplay.org/

Sesame Workshop
http://www.sesameworkshop.org/

Sing, Talk, Read (STAR)
http://www.learndc.org/earlychildhood/sing-talk-read

Talk, Read, Play (Boston, MA)
http://www.talkreadplay.org/

Talk With Me Baby (Georgia)
http://www.talkwithmebaby.org/

Text4baby
https://www.text4baby.org/

Too Small to Fail's "Talking Is Teaching: Talk, Read, Sing" Campaign
http://www.toosmall.org

http://www.talkingisteaching.org

Univision and Too Small to Fail's "Pequeños y Valiosos" Campaign
http://noticias.univision.com/educacion/pequenos-y-valiosos

Vroom

> http://www.joinvroom.org/

Zero to Three

> http://www.zerotothree.org/

American Speech-Language-Hearing Association (ASHA)

> http://www.asha.org/

Policy Centers And Resources

Brookings's Social Policy: Early Childhood Development

> http://www.brookings.edu/research/topics/early-child
> -development

Center for Law and Social Policy

> http://www.clasp.org/

Committee for Economic Development

> www.ced.org

Division for Early Childhood

> http://www.dec-sped.org/

Heckman Equation

> http://www.heckmanequation.org

Kids Count, Annie E. Casey Foundation

> http://www.aecf.org/work/kids-count/

National Academy for State Health Policy

> www.nashp.org

National Center for Children in Poverty

> www.nccp.org

National Women's Law Center

> http://www.nwlc.org/our-issues/child-care-%2526-early
> -learning/head-start

New America Foundation
   http://earlyed.newamerica.net/dashboard

Office of the Administration for Children and Families
   http://www.acf.hhs.gov/programs/ecd

Policy Report: Creating Opportunities for Families: A Two-
   Generation Approach
   http://www.aecf.org/m/resourcedoc/aecf
   -CreatingOpportunityforFamilies-2014.pdf#page=3

Report of the National Early Childhood Accountability Task Force
   http://policyforchildren.org/wp-content/uploads/2013/07
   /Taking-Stock.pdf

U.S. Department of Education: Early Learning Initiative
   http://www.ed.gov/early-learning

Research Centers And Projects

Center for Children and Families: University of Texas at Dallas
   http://ccf.utdallas.edu/

Center for the Economics of Human Development
   http://www.cchd.uchicago.edu

Center on the Developing Child: Harvard University
   http://developingchild.harvard.edu/

Center for Early Education and Development: University of Minnesota
   http://www.cehd.umn.edu/ceed/

Child Trends
   www.childtrends.org

Columbia University: National Center for Children and Families
   http://policyforchildren.org/research-projects/early-care
   -education/

Erikson Institute Herr Research Center for Children and Social Policy
http://www.erikson.edu/herr-research-center/

Frank Porter Graham Child Development Institute
http://fpg.unc.edu/

Proyecto Habla conmigo! (Talk with Me!): Language Learning Lab,
Stanford University
http://web.stanford.edu/group/langlearninglab/cgi-bin/habla
conmigo.php

Harvard Family Research Project: The Family Involvement Network
of Educators (FINE)
http://www.hfrp.org/family-involvement/fine-family
-involvement-network-of-educators

Human Early Learning Partnership
http://earlylearning.ubc.ca/

Institute for Learning and Brain Sciences (I-LABS)
http://ilabs.uw.edu/

Juniper Gardens Children's Project
http://jgcp.ku.edu/

KidTalk
http://vkc.mc.vanderbilt.edu/kidtalk/

Mind in the Making
http://www.mindinthemaking.org/

Thirty Million Words Initiative
http://thirtymillionwords.org/

University of Chicago: Center for Early Childhood Research
http://babylab.uchicago.edu/page/our-research

What Works Clearinghouse: Institute of Education Sciences
http://ies.ed.gov/ncee/wwc/topic.aspx?sid=4

International Programs

1,000 Days
　　http://www.thousanddays.org/

The Hanen Centre
　　http://www.hanen.org/Home.aspx

Oxford's Handbook of Early Childhood Development Research
　　http://www.oxfordscholarship.com/view/10.1093/acprof:
　　oso/9780199922994.001.0001/acprof-9780199922994

UNICEF Early Childhood Development
　　http://www.unicef.org/earlychildhood/

WHO's Commission on the Social Determinants of Health: Early
　　Childhood Development
　　http://whqlibdoc.who.int/hq/2007/a91213.pdf?ua=1

It should be noted that, while all those described in the book are real, and their stories true, names have been changed to protect their privacy.

Chapter 1: Connections

4   *before leaving the hospital:* National Institutes of Health, *Fact sheet: Newborn hearing screening,* National Institute on Deafness and Other Communication Disorders, 2010, accessed December 16, 2014, http:// report.nih.gov/nihfactsheets/Pdfs/NewbornHearingScreening(NIDCD) .pdf.

5   *approved for young children:* National Institutes of Health, *Fact sheet: Cochlear implants,* National Institute on Deafness and Other Communication Disorders, 2010, accessed December 16, 2014, http://report .nih.gov/nihfactsheets/Pdfs/CochlearImplants(NIDCD).pdf.

5   *negative impact on a newborn's development:* Dimity Dornan, "Hearing loss in babies is a neurological emergency," *Alexander Graham Bell Association for the Deaf and Hard of Hearing* (2009), accessed December 17, 2014, http://www.hearandsayresearchandinnovation.com.au /UserFiles/files/Publications/Dornan%202009%20Hearing%20loss %20emergency.pdf.

13   *one-third of deaf adults are functionally illiterate:* Connie Mayer, "What really matters in the early literacy development of deaf children," *Journal of Deaf Studies and Deaf Education* 12.4 (2007): 411–431. Statistics quoted from p. 412.

13   *predictor of long-term academic success:* Joy Lesnick, Robert M. Goerge, and Cheryl Smithgall, "Reading on grade level in third grade:

How is it related to high school performance and college enrollment," *Chicago: Chapin Hall at the University of Chicago* (2010).

13   *Nim Tottenham:* "Parents are playing a really big role in shaping children's brain development." And parenting, she says, is a bit like oxygen. It's easy to take for granted until you see someone who isn't getting enough. Statement attributed to Nim Tottenham, in Jon Hamilton, "Orphans' lonely beginnings reveal how parents shape a child's brain," *Shots: Health News from National Public Radio*, National Public Radio, February 24, 2014.

22   *stark disparities in ultimate educational achievement:* Betty Hart and Todd R. Risley, *Meaningful Differences in the Everyday Experience of Young American Children* (Baltimore: Paul H. Brookes, 1995).

Chapter 2: The First Word

26   *"cumulative vocabulary growth curves":* Betty Hart and Todd Risley, "The early catastrophe: The 20 million word gap by age 3," *American Education* (Spring 2003): 1, accessed December 19, 2014, http://www.aft.org/sites/default/files/periodicals/TheEarlyCatastrophe .pdf.

26   *had not attended the intensive preschool program:* Betty Hart and Todd Risley, *Meaningful Differences.* See chap. 1, "Intergenerational transmission of competence," pp. 1–20, for a full discussion of Hart and Risley's first interventions and their results.

26   *break the "cycle of poverty" through preschool education:* T.R. Risley and B. Hart, "Promoting early language development," in *The Crisis in Youth Mental Health: Critical Issues and Effective Programs,* vol. 4, *Early Intervention Programs and Policies,* pp. 83–88, ed. N.F. Watt, C. Ayoub, R.H. Badley, J.E. Puma, and W.A. LeBoef (Westport, CT: Praeger, 2006); also presented as an open source article: Todd R. Risley, "The everyday experience of American babies: Discoveries and implications," *Senior Dad,* accessed January 8, 2015, http://srdad.com /SrDad/Early_Childhood_files/Todd%20Risley.pdf.

26   *"to cure it and, above all, to prevent it":* "President Lyndon B. Johnson's Annual Message to the Congress on the State of the Union," January 8, 1964 (delivered in person before a joint session), *LBJ Presidential Library,* accessed December 19, 2014, http://www.lbjlib .utexas.edu/johnson/archives.hom/speeches.hom/640108.asp.

26   *in the basement of C.L. Davis's liquor store:* R.V. Hall, R.L. Schiefelbuch, R.K. Hoyt, and C.R. Greenwod, "History, mission and organi-

zation of the Juniper Gardens Children's Project," *Education and Treatment of Children*, 12.4 (1989): 301–329. Discussion of C.L. Davis Liquor Store on p. 306.

26   *children's school readiness and academic potential:* "Spearhead—Juniper Gardens Children's Project," YouTube video, 7:04, posted by "JuniperGardensKU," January 30, 2013, accessed December 19, 2014, http://www.youtube.com/watch?v=bW77QiceqOE.

26   *YouTube video:* Ibid.

27   *described Betty Hart and Todd Risley as "romantics":* Steve Warren, telephone interview with the author, February 20, 2014.

28   *"shocking when you think about it":* Todd R. Risley, interview by David Bouton, December 14, 2004, transcript, *Children of the Code*, accessed December 19, 2014, http://www.childrenofthecode.org/interviews/risley.htm.

28   *Skinner's "operant conditioning":* Marc N. Branch, "Operant conditioning," in *Encyclopedia of Human Development*, ed. Neil J. Salkind (Thousand Oaks, CA: Sage Publications, 2005), accessed December 19, 2014, http://www.sage-ereference.com/view/humandevelopment/n458.xml?rskey=t8Ib4Landrow=5.

29   *Chomsky theorized:* C.f. J. Michael Bowers, "Language acquisition device," in *Encyclopedia of Human Development*, ed. Neil J. Salkind (Thousand Oaks, CA: Sage Publications, 2005), accessed December 19, 2014, http://www.sage-ereference.com/view/humandevelopment/n371.xml.

29   *Dismissing B.F. Skinner's hypothesis as "absurd":* Noam Chomsky, "Review: Verbal behavior by B. F. Skinner," *Linguistic Society of America* 35.1 (1959): 26–58.

29   *disparities in language outcomes were rare:* For a discussion of the impact of Noam Chomsky's theory of universal grammar on research related to social disparities in earl language acquisition, see p. 8 in A. Fernald and A. Weisleder, "Early language experience is vital to developing fluency in understanding," in *Handbook of Early Literacy Research*, vol. 3, ed. S. Neuman and D. Dickinson (New York: Guildford Publications, 2011), pp. 2–20. See also p. 184 in A. Fernald and V.A. Marcham, "Causes and consequences of variability in early language learning," in *Experience, Variation and Generalization: Learning a First Language*, ed. I. Arnon and E.V. Clark (Philadelphia: John Benjamins, 2011), pp. 181–202.

29   *examine variations in development:* Fernald and Marcham, "Causes and consequences." See p. 185 for a discussion of the problematic as-

sumption that the developmental patterns observed in middle-class children are generalizable to children from all walks of life.

30 *Do Good And Take Data:* Glen Dunlap and John R. Lutzker, "Todd R. Risley (1937–2007)," *Journal of Positive Behaviour Intervention* 10.9 (2008): 148–149, accessed January 8, 2015, http://pbi.sagepub.com /content/10/3/148.full.pdf+html. Quote on p. 148.

30 *"develop answers to serious human problems":* Ibid.

30 *"to the essence of the problem":* James A. Sherman, "Todd R. Risley: Friend, colleague, visionary," *Journal of Applied Behavior Analysis* 41.1 (2008): 7–10. Quote on p. 9.

30 *"a unique genius":* Steve Warren, telephone interview with the author, February 20, 2014.

31 *nine months of age to three years:* B. Hart and T.R. Risley, "American parenting of language-learning in children: Persisting differences in family-child interactions observed in natural home environments," *Developmental Psychology* 28 (1992): 1096–1105.

31 *stay in one place for the foreseeable future:* For detailed discussion, see chap. 3, "42 American Families," in Hart and Risley, *Meaningful Differences,* pp. 53–74.

32 *"the more we could potentially learn":* Ibid., p. 24.

32 *everything "done by the children, to them, and around them":* Ibid., p. 24 (emphasis added).

32 *a single day of vacation during the entire study:* Ibid., p. 41.

32 *"finally ready to begin formulating what it all meant":* Ibid., p. 46.

32 *twenty thousand work hours, analyzing the data, seems almost unbelieveable:* Todd R. Risley and Betty Hart, "Promoting early language development," in *The Crisis in Youth Mental Health: Critical Issues and Effective Programs,* vol. 4, *Early Intervention Programs and Policies,* ed. N.F. Watt, C. Ayoub, R.H. Bradley, J.E. Puma, and W.A. LeBoeuf (Westport, CT: Praeger, 2006), pp. 83–88, as cited in Risley, "Everyday experiences of American babies."

32 *referred to Betty Hart as a "foreman":* Risley, interview.

33 *"see all the other children [doing the same thing]":* Hart and Risley, *Meaningful Differences,* p. 54.

33 *"Do you have to go potty?":* Ibid., pp. 53–54.

33 *"conditions for language learning":* Ibid., p. 55.

33 *"basic skills needed for preschool entry":* Hart and Risley, "Early catastrophe," p. 7.

33 *"frequent periods of silence":* Hart and Risley, *Meaningful Differences,* p. 60.

33   *less than half that:* Ibid., pp. 64–66.

34   *children of welfare families heard about six* hundred: Ibid., p. 132.

34   *fewer than 50 times in the same period:* Ibid., pp. xx and 124, fig. 9.

34   *Children in welfare homes, about four:* Ibid., pp. 126 and 128, figs. 11 and 12.

35   *From Thirteen Months to Thirty-six Months:* Ibid., pp. 66 and 176, tab. 5.

35   *Extrapolating to One Year:* Ibid., p. 71; Hart and Risley, "Early catastrophe," p. 8.

35   *The Staggering Difference:* Hart and Risley, *Meaningful Differences,* pp. 197–198; Hart and Risley, "Early catastrophe," p. 8.

35   *Thirty Million Words:* Hart and Risley, *Meaningful Differences,* pp. 197–198; see also p. 198, fig. 19.

35   *Three Years of Age:* Hart and Risley, "Early catastrophe," p. 7.

36   *effect on IQ at three years of age:* Hart and Risley, *Meaningful Differences,* pp. 143–144.

36   *"I.Q. test scores at age three and later":* Ibid., pp. xx and 144.

37   *"'Don't' 'Stop' 'Quit that'":* Ibid., p. 147.

37   *"talk and behave like their families":* Ibid., p. 58.

37   *[was identical to what they heard at home]:* Ibid., p. 59.

37   *school test scores at ages nine and ten:* Dale Walker, Charles Greenwood, Betty Hart, and Judith Carta, "Prediction of school outcomes based on early language production and socioeconomic factors," *Child Development* 65 (1994): 606–621.

38   *Flávio Cunha provided the following assessment of the Hart and Risley study:* The following discussion is based on personal communication from Flávio Cunha on May 18, 2014.

40   *"people and circumstances they did not include":* Hart and Risley, "Early catastrophe," p. 8.

40   *"the truly disadvantaged":* William Julius Wilson, *The Truly Disadvantaged* (Chicago: University of Chicago Press, 2012).

40   *"in public housing with single mothers . . . in virtual silence":* S.B. Heath, "The children of Trackton's children: Spoken and written language in social change," in *Cultural Psychology: Essays on Comparative Human Development,* ed. J.W. Stilger, R.A. Shweder, and G. Herdt (Cambridge: Cambridge University Press, 1990), pp. 496–519, cited in E. Hoff, "How social contexts support and shape language development," *Developmental Review* 26 (2006): 55–88, p. 60.

41   *no matter their socioeconomic status:* See Hart and Risley, *Meaningful*

Differences, chap. 6, "The early experience of 42 typical American children," pp. 119–140, for a detailed discussion of the relationship between the quantity and quality of child-directed parent talk.

41 *"We just have to help them [talk] more":* Risley, "Everyday experience of American babies," p. 3.

42 *the icing on the cake:* Ibid.

43 *Lower-income parents, on the other hand, began, then stopped:* Hart and Risley, *Meaningful Differences,* pp. 124–125.

44 *children in lower socioeconomic status families:* Ibid., pp. 125–126.

44 *five times more than welfare children heard:* Ibid., p. 126.

44 *In One Year:* Hart and Risley, "Early catastrophe," p. 8.

45 *Four Years Of Age:* See fig. 20 in Hart and Risley, *Meaningful Differences,* pp. 200 and 253; p. 253 contains more details of their analysis.

46 *"help [these students] overcome all obstacles":* Shayne Evans, personal communication, June 9, 2014.

48 *It is a process critical to learning:* Anne Fernald, "Why efficiency in processing language is important," YouTube video, 2:24, posted by "Treeincement," June 11, 2010, accessed December 19, 2014, https://www.youtube.com/watch?v=verqCmPrnY8.

48 *then all that follows is lost, as well:* Ibid.

48 *"buys you the opportunity to learn":* Ibid.

Chapter 3: Neuroplasticity

52 *Anne Fernald's elegant study:* Anne Fernald, Virginia A. Marchman, and Adriana Weisleder, "SES differences in language processing skill and vocabulary are evident at 18 months," *Developmental Science* 16.2 (2013): 234–248.

53 *Jack Shonkoff:* National Scientific Council on the Developing Child, "The timing and quality of early experiences combine to shape brain architecture (working paper 5, Center on the Developing Child at Harvard University, Cambridge, MA, 2007), quote on p. 2, accessed January 9, 2015, http://www.developingchild.net.

54 *more constructive, less negatively reactive ways:* "Toxic Stress Derails Healthy Development," Center on the Developing Child at Harvard University video, 1:53, 2014, accessed January 9, 2015, http://developingchild.harvard.edu/resources/multimedia/videos/three_core_concepts/toxic_stress/.

55 *a lifetime of learning, behavior, and health is built:* National Scientific Council on the Developing Child, "Timing and quality."

55 *the unforgettable "still face" experiment on YouTube:* Edward Tron-

ick, "Still face experiment: Dr. Edward Tronick," YouTube video, 2:49, posted by "UMass Boston," November 30, 2009, accessed January 14, 2015, https://www.youtube.com/watch?v=apzXGEbZhto.

57   *permanently affecting the developing brain:* National Scientific Council on the Developing Child, "Timing and quality," p. 8.

57   *memory, emotion, behavior, motor skills, and, of course, language:* "Five numbers to remember about early childhood development," Center on the Developing Child at Harvard University, 2014, accessed January 9, 2015, http://developingchild.harvard.edu/resources/multimedia /interactive_features/five-numbers/.

58   *learning a new language as you get older, increasingly difficult:* National Scientific Council on the Developing Child, "Timing and quality," pp. 2–3.

61   *revolutionized our understanding of brain plasticity:* Martha Constantine-Paton, "Pioneers of cortical plasticity: Six classic papers by Wiesel and Hubel," *Journal of Neurophysiology* 99.6 (2008): 2741–2744; Joel Davis, "Brain and visual perception: The story of a 25-year collaboration," *Color Research and Application* (2005): 3.

61   *"one of the most well-guarded secrets of the brain":* Alyssa A. Botelho, "David H. Hubel, Nobel Prize-winning neuroscientist, dies at 87," *The Washington Post,* September 23, 2013, accessed January 9, 2015, http://www.washingtonpost.com/local/obituaries/david-h-hubel-nobel -prize-winning-neuroscientist-dies-at-87/2013/09/23/5a227c2c-7167-11e2 -ac36-3d8d9dcaa2e2_story.html.

62   *included the two of them dancing and sexy pictures of women:* Ibid.

62   *changed the way we understand the brain:* Botelho, "David H. Hubel."

62   *"It only helps to explain the workings of the mind":* Eric R. Kandel, "An introduction to the work of David Hubel and Torsten Wiesel," *Journal of Physiology* 587.12 (2009): 2733–2741; quote on p. 2733.

64   *essential key to avoiding that cost:* Liz Schroeder, S. Petrou, C. Kennedy, D. McCann, C. Law, P.M. Watkin, S. Worsfold, and H.M. Yuen, "The economic costs of congenital bilateral permanent childhood hearing impairment," *Pediatrics* 117.4 (2006): 1101–1112.

66   *without ever having actually heard English or knowing how it sounds:* Charlene Chamberlain, Jill P. Morford, and Rachel I. Mayberry, eds., *Language Acquisition by Eye* (Mahwah, NJ: Lawrence Erlbaum Associates, 2000).

66   *predicts high school graduation:* Keith E. Stanovich, "Matthew effects in reading: Some consequences of individual differences in the acquisition of literacy," *Reading Research Quarterly* (1986): 360–407.

66   *significantly less than for a hearing child:* Harry G. Lang, "Higher

education for deaf students: Research priorities in the new millennium," *Journal of Deaf Studies and Deaf Education* 7.4 (2002): 267–280. Statistic for postsecondary graduation of the deaf in the United States on p. 268.

66 *45 percent less than their hearing counterparts:* C. Reilly and S. Qi, "Snapshot of deaf and hard of hearing people, postsecondary attendance and unemployment," 2011, accessed December 18, 2014, http://research.gallaudet.edu/Demographics/deaf-employment-2011.pdf; Bonnie B. Blanchfield, Jacob J. Feldman, Jennifer L. Dunbar, and Eric N. Gardner, "The severely to profoundly hearing-impaired population in the United States: prevalence estimates and demographics," *Journal of the American Academy of Audiology* 12.4 (2001): 183–189; John M. McNeil, "Employment, earnings, and disability" (paper prepared for the 75th Annual Conference of the Western Economic Association, Vancouver, BC, June 29–July 3, 2000), accessed December 18, 2014, http://www.vocecon.com/resources/ftp/Bibliography/mcnempl.pdf.

68 *Patricia Kuhl:* Marcie Sillman, "Brain waves: Peeking under the hood," *KUOW News,* radio, Washington, 2010. January 14, 2015.

68 *"a hair dryer from Mars":* Patricia Kuhl quoted in Meeri Kim, "Babies grasp speech before they utter first word, a study finds," *The Washington Post,* July 19, 2014, accessed January 14, 2015, http://www.washingtonpost.com/national/health-science/babies-grasp-speech-before-they-utter-their-first-word-a-study-finds/2014/07/19/c4854b46-0ea8-11e4-8c9a-923eccoc7d23_story.html.

68 *"computational geniuses":* Patricia Kuhl, "The first year 'computational geniuses,'" *National Geographic,* 2015, accessed January 14, 2015, http://ngm.nationalgeographic.com/2015/01/baby-brains/geniuses-video.

69 *the process may actually begin in utero:* Christine Moon, Hugo Lagercrantz, and Patricia K. Kuhl, "Language experienced in utero affects vowel perception after birth: A two-country study," *Acta Paediatrica* 102.2 (2013): 156–160.

69 *"an articulate seven-year-old with a mouth full of marbles":* Isaac Stone Fish, "Mark Zuckerberg speaks Mandarin like a seven-year-old," *Foreign Policy,* Passport, October 22, 2014, accessed January 14, 2015, http://foreignpolicy.com/2014/10/22/mark-zuckerberg-speaks-mandarin-like-a-seven-year-old/.

69 *rather than one billion:* David Goldman and Sophia Yan, "Zuckerberg, in all-Chinese Q&A, says Facebook has '11 mobile users.'" CNN Money, October 23, 2014, accessed January 14, 2015, http://money.cnn.com/2014/10/23/technology/social/zuckerberg-chinese/index.html?hpt=ob_articlefooterandiid=obnetwork.

70   *Babies are true "citizens of the world":* Patricia Kuhl, "The linguistic genius of babies," filmed October 2010, TED video, 10:17, accessed January 14, 2015, https://www.ted.com/speakers/patricia_kuhl.

71   *even those with slight variations:* This is known as the native language magnet theory. For a detailed discussion, see Patricia K. Kuhl, "Psychoacoustics and speech perception: Internal standards, perceptual anchors, and prototypes," in *Developmental Psychoacoustics,* ed. Lynne A. Werner and Edwin W. Rubel (Washington, DC: American Psychological Association, 1992); Patricia K. Kuhl, "Learning and representation in speech and language," *Current Opinion in Neurobiology* 4.6 (1994): 812–822. For an updated and elaborated version of this theory, see Patricia K. Kuhl, Barbara T. Conboy, Sharon Coffey-Corina, Denise Padden, Maritza Rivera-Gaxiola, and Tobey Nelson, "Phonetic learning as a pathway to language: New data and native language magnet theory expanded (NLM-e)," *Philosophical Transactions of the Royal Society B: Biological Sciences* 363.1493 (2008): 979–1000.

71   *three months later, the ability had disappeared:* For a discussion of this research, see Alison Gopnik, Andrew N. Meltzoff, and Patricia K. Kuhl, *The Scientist in the Crib: Minds, Brains, and How Children Learn* (New York: Harper, 1999), pp. 104–110.

72   *connect humans to other humans:* For detailed discussions of the social-evolutionary nature of language learning, see Ralph Adolphs, "Cognitive neuroscience of human social behaviour," *Nature Reviews Neuroscience* 4.3 (2003): 165–178; Robin I.M. Dunbar, "The social brain hypothesis," *Evolutionary Anthropology* 6 (1998): 178–190; Friedemann Pulvermüller, "Brain mechanisms linking language and action," *Nature Reviews Neuroscience* 6.7 (2005): 576–582.

73   *its importance cannot be emphasized enough:* For research related to how the brain hears and understands meaningful sound segments, see Allison J. Doupe and Patricia K. Kuhl, "Birdsong and human speech: common themes and mechanisms," *Annual Review of Neuroscience* 22.1 (1999): 567–631; Cristopher S. Evans and Peter Marler, "Language and animal communication: Parallels and contrasts," in *Comparative Approaches to Cognitive Science,* ed. Herbert L. Roitblat and Jean-Arcady Meyer (Cambridge, MA: MIT Press, 1995), pp. 341–382; Peter Marler, "Song-learning behavior: the interface with neuroethology," *Trends in Neurosciences* 14.5 (1991): 199–206.

73   *You guessed it. Nothing:* For a clear summary of this research, see Kuhl, "Linguistic genius of babies."

74   *putting an end date on the plasticity of the brain:* For a good overview of

Takao Hensch's research findings, see Jon Bardin, "Neurodevelopment: Unlocking the brain," *Nature* 487.7405 (2012): 24–26, accessed January 16, 2015, http://www.nature.com/news/neurodevelopment-unlocking -the-brain-1.10925.

74   *supposed to be lost in early childhood:* Ibid.

75   *"make up for lost time is compelling":* Ibid.

Chapter 4: The Power of Parent Talk

79   *New York and the wiring of the brain are very similar:* Jon Hamilton, "How your brain is like Manhattan," *Shots: Health News from NPR,* National Public Radio, March 29, 2012, accessed January 15, 2015, http://www.npr.org/blogs/health/2012/03/29/149629657/how-your-brain -is-like-manhattan.

79   *how we think and how we behave:* Sebastian Seung, *Connectome: How the Brain's Wiring Makes Us Who We Are* (Boston: Houghton Mifflin Harcourt, 2012), xv.

80   *only open up new questions:* James Gorman, "Learning how little we know about the brain," *The New York Times,* Science section, November 10, 2014, accessed January 15, 2015, http://www.nytimes.com/2014/11/11 /science/learning-how-little-we-know-about-the-brain.html?_r=o.

80   *"nature meeting nurture":* Sebastian Seung, "I am my connectome," filmed February 2010, TED video, 19:25, accessed January 15, 2015, https://www.ted.com/talks/sebastian_seung.

83   *funny and not so funny:* Elizabeth Green, "Why do Americans stink at math?" *The New York Times,* July 23, 2014, accessed January 6, 2015, http://www.nytimes.com/2014/07/27/magazine/why-do-americans -stink-at-math.html?_r=o.

83   *a fourth of a pound at McDonald's:* A. Alfred Taubman, *Threshold Resistance: The Extraordinary Career of a Luxury Retailing Pioneer* (New York: Harper Business, 2007).

84   *"Taking the math out of medicine":* Home page, eBroselow.com, 2015, accessed January 6, 2015, www.eBroselow.com.

84   *in 2012 the United States ranked:* Organisation for Economic Co-operation and Development, "Programme for International Student Assessment (PISA) Results from PISA 2012," Country Note: United States, 2012, accessed January 15, 2015, http://www.oecd.org/pisa /keyfindings/PISA-2012-results-US.pdf.

86   *"strong-performing U.S. states":* Ibid.

86   *more than 16 percent in Canada:* Stephanie Simon, "PISA results show 'educational stagnation' in U.S.," *Politico,* December 3, 2013, accessed

January 15, 2015, http://www.politico.com/story/2013/12/education
-international-test-results-100575.html#ixzz3JonddFu.

86   *first grade and even kindergarten:* Jae H. Paik, Loes van Gelderen,
Manuel Gonzales, Peter F. de Jong, and Michael Hayes, "Cultural differences in early math skills among US, Taiwanese, Dutch, and Peruvian preschoolers," *International Journal of Early Years Education*
19.2 (2011): 133–143.

86   *similar to U.S. second graders for number estimation:* See, for example, the following studies cited in Paik et al., "Cultural differences in
early math skills"; David C. Geary, Liu Fan, and C. Christine Bow-
Thomas, "Numerical cognition: Loci of ability differences comparing
children from China and the United States," *Psychological Science* 3.3
(1992): 180–185; Robert S. Siegler and Julie L. Booth, "Development
of numerical estimation in young children," *Child Development* 75.2
(2004): 428–444; Robert S. Siegler and Yan Mu, "Chinese children
excel on novel mathematics problems even before elementary school,"
Psychological Science 19.8 (2008): 759–763.

87   *logical next step from ten:* Kevin Miller, Susan M. Major, Hua Shu,
and Houcan Zhang, "Ordinal knowledge: Number names and number
concepts in Chinese and English," *Canadian Journal of Experimental
Psychology/Revue canadienne de psychologie expérimentale* 54.2
(2000): 129–140.

87   *appears to be notably different:* Xin Zhou, Jin Huang, Zhengke Wang,
Bin Wang, Zhenguo Zhao, Lei Yang, and Zhengzheng Yang, "Parent-
child interaction and children's number learning," *Early Child Development and Care* 176.7 (2006): 763–775.

87   *ready to absorb math according to individual innate ability:* Prentice
Starkey and Alice Klein, "Sociocultural influences on young children's
mathematical knowledge," *Contemporary Perspectives on Mathematics in Early Childhood Education* (2008): 253–276.

87   *not ready for abstract mathematical thinking:* David P. Weikart, *The
Cognitively Oriented Curriculum: A Framework for Preschool Teachers* (Washington, DC: National Association for the Education of Young
Children, 1971), cited in Starkey and Klein, "Sociocultural influences
on young children's mathematical knowledge."

88   *"not a glimmering . . . idea of numbers":* Richard W. Copeland, *How
Children Learn Mathematics: Teaching Implications of Piaget's Research* (New York: Macmillan, 1970), 374, cited in Starkey and Klein,
"Sociocultural influences on young children's mathematical knowledge."

88   *correct number of geometric shapes:* Véronique Izard, Coralie Sann,

Elizabeth S. Spelke, and Arlette Streri, "Newborn infants perceive abstract numbers," *Proceedings of the National Academy of Sciences* 106.25 (2009): 10382–10385.

88 *often predicted their eventual math ability:* Ariel Starr, Melissa E. Libertus, and Elizabeth M. Brannon, "Number sense in infancy predicts mathematical abilities in childhood," *Proceedings of the National Academy of Sciences* 110.45 (2013): 18116–18120.

88 *basic mathematical procedures related to those estimates:* Hilary Barth, Kristen La Mont, Jennifer Lipton, and Elizabeth S. Spelke, "Abstract number and arithmetic in preschool children," *Proceedings of the National Academy of Sciences of the United States of America* 102.39 (2005): 14116–14121; Camilla K. Gilmore, Shannon E. McCarthy, and Elizabeth S. Spelke, "Non-symbolic arithmetic abilities and mathematics achievement in the first year of formal schooling," *Cognition* 115.3 (2010): 394–406; Koleen McCrink and Elizabeth S. Spelke, "Core multiplication in childhood," *Cognition* 116.2 (2010): 204–216; Koleen McCrink and Karen Wynn, "Large-number addition and subtraction by 9-month-old infants," *Psychological Science* 15.11 (2004): 776–781.

91 *literacy achievement through the third grade:* Greg J. Duncan, C.J. Dowsett, A. Claessens, K. Magnuson, A.C. Huston, P. Klebanov, L.S. Pagani, et al., "School readiness and later achievement," *Developmental Psychology* 43.6 (2007): 1428–1446.

91 *math to age fifteen:* Tyler W. Watts, Greg J. Duncan, Robert S. Siegler, and Pamela E. Davis-Kean, "What's past is prologue: Relations between early mathematics knowledge and high school achievement," *Educational Researcher* 43.7 (2014): 352–360.

92 *could be contrasted to almost 100,000 for others:* Susan C. Levine, Linda Whealton Suriyakham, Meredith L. Rowe, Janellen Huttenlocher, and Elizabeth A. Gunderson, "What counts in the development of young children's number knowledge?," *Developmental Psychology* 46.5 (2010): 1309–1319.

93 *spatial intelligence allowed genius to compound genius:* Aaron Klug, "From macromolecules to biological assemblies" (Nobel lecture, December 8, 1982), accessed January 15, 2015, http://www.nobelprize .org/nobel_prizes/chemistry/laureates/1982/klug-lecture.pdf.

94 *entirely related to their experience with spatial words:* William Harms, "Learning spatial terms improves children's spatial skills," *UChicago-News,* November 9, 2011, accessed January 15, 2015, http://news.uchicago .edu/article/2011/11/09/learning-spatial-terms-improves-childrens-spatial -skills.

95 *received half the overall math talk than did sons:* Alicia Chang, Catherine M. Sandhofer, and Christia S. Brown, "Gender biases in early number exposure to preschool-aged children," *Journal of Language and Social Psychology* 30.4 (2011): 440–450.

95 *important STEM fields of science, technology, engineering, and math:* Rebecca Carr, "Women in the Academic Pipeline for Science, Technology, Engineering and Math: Nationally and at AAUDE Institutions," Association of American Universities Data Exchange, April 2013, accessed February 2, 2015 http://aaude.org/system/files/documents/public /reports/report-2013-pipeline.pdf.

96 *girls as young as seven:* Pascal Huguet and Isabelle Régner, "Counter-stereotypic beliefs in math do not protect school girls from stereotype threat," *Journal of Experimental Social Psychology* 45.4 (2009): 1024–1027; Emmanuelle Neuville and Jean-Claude Croizet, "Can salience of gender identity impair math performance among 7–8 years old girls? The moderating role of task difficulty," *European Journal of Psychology of Education* 22.3 (2007): 307–316.

96 *women going into math, engineering, and computer science:* For detailed discussions of how stereotype threat affects performance in STEM subjects, see Albert Bandura, Claudio Barbaranelli, Gian Vittorio Caprara, and Concetta Pastorelli, "Self-efficacy beliefs as shapers of children's aspirations and career trajectories," *Child Development* 72.1 (2001): 187–206; Carol Dweck, *Mindset: The New Psychology of Success* (New York: Ballantine Books, 2006); Peter Häussler and Lore Hoffmann, "An intervention study to enhance girls' interest, self-concept, and achievement in physics classes," *Journal of Research in Science Teaching* 39.9 (2002): 870–888.

96 *girls do as well as boys in math:* Janet S. Hyde, Sara M. Lindberg, Marcia C. Linn, Amy B. Ellis, and Caroline C. Williams, "Gender similarities characterize math performance," *Science* 321.5888 (2008): 494–495; Janet S. Hyde and Janet E. Mertz, "Gender, culture, and mathematics performance," *Proceedings of the National Academy of Sciences* 106.22 (2009): 8801–8807.

96 *females working in the STEM fields is also increasing:* Stephen J. Ceci, Donna K. Ginther, Shulamit Kahn, and Wendy M. Williams, "Women in academic science: A changing landscape," *Psychological Science in the Public Interest* 15.3 (2014): 75–141.

96 *confronted with actual math achievement:* Pamela M. Frome and Jacquelynne S. Eccles, "Parents' influence on children's achievement-related perceptions," *Journal of Personality and Social Psychology* 74.2 (1998): 435–452.

96   *influencing both actual participation and interest:* Sandra D. Simp-kins, Pamela E. Davis-Kean, and Jacquelynne S. Eccles, "Math and science motivation: A longitudinal examination of the links between choices and beliefs," *Developmental Psychology* 42.1 (2006): 70–83, http://dx.doi.org/10.1037/0012-1649.42.1.70.

96   *successful in math-related careers:* Martha M. Bleeker and Janis E. Jacobs, "Achievement in math and science: Do mothers' beliefs matter 12 years later?," *Journal of Educational Psychology* 96.1 (2004): 97–109.

97   *no matter how well they're actually doing in math:* Jennifer Herbert and Deborah Stipek, "The emergence of gender differences in children's perceptions of their academic competence," *Journal of Applied Developmental Psychology* 26.3 (2005): 276–295.

97   *Professor Sian Beilock, the author of* Choke: Sian Beilock, *Choke: What the Secrets of the Brain Reveal About Getting It Right When You Have To* (New York: Free Press, 2010).

97   *elementary school teachers on math achievement:* Sian L. Beilock, Elizabeth A. Gunderson, Gerardo Ramirez, and Susan C. Levine, "Female teachers' math anxiety affects girls' math achievement," *Proceedings of the National Academy of Sciences* 107.5 (2010): 1860–1863.

97   *unrelated to their teacher's level of "math anxiety":* Ibid.

98   *more likely not to exhibit gender stereotyping in math:* Ibid.

99   Mindset: The Psychology of Success: Dweck, *Mindset.*

101   *"marked changes in their motivation to learn":* Ibid., p. 51.

102   The Psychology of Self-Esteem: Nathaniel Branden, *The Psychology of Self-Esteem: A Revolutionary Approach to Self-Understanding That Launched a New Era in Modern Psychology* (San Francisco: Jossey-Bass, 1969).

102   *John Vasconcellos:* California State Department of Education, Sacramento, "Toward a state of esteem: The final report of the California task force to promote self-esteem and personal and social responsibility" (1990).

102   *"Toward a State of Esteem":* Ibid.

103   *"dependent on the opinion of others":* Carol S. Dweck, "Caution—praise can be dangerous," *American Educator* 23.1 (1999): 4–9.

105   *praised for working hard chose the more difficult task:* See study results in Claudia M. Mueller and Carol S. Dweck, "Praise for intelligence can undermine children's motivation and performance," *Journal of Personality and Social Psychology* 75.1 (1998): 33–52; p. 36.

106   *versus praise for effort:* Elizabeth A. Gunderson, Sarah Gripshover, Carissa Romero, Carol S. Dweck, Susan Goldin-Meadow, and Susan

Cohen Levine, "Parent praise to 1- to 3-year-olds predicts children's motivational frameworks 5 years later," *Child Development* 84.5 (2013): 1526–1541.

106 *achievement from second to fourth grades:* S. Gripshover, N. Sorhagen, E.A. Gunderson, C.S. Dweck, S. Goldin-Meadow, and S.C. Levine, "Parent praise to toddlers predicts fourth grade academic achievement via children's incremental mindsets" (manuscript under review).

109 *higher grade point average than the control group:* Geoffrey L. Cohen, Julio Garcia, Nancy Apfel, and Allison Master, "Reducing the racial achievement gap: A social-psychological intervention," *Science* 313.5791 (2006): 1307–1310.

111 *more likely, years later, to do better academically:* Professor Mischel's work is described in detail in his book: Walter Mischel, *The Marshmallow Test: Mastering Self-control* (New York: Little, Brown, 2014).

112 *its draining complexities, is stressful:* Clancy Blair, "Stress and the development of self-regulation in context," *Child Development Perspectives* 4.3 (2010): 181–188; National Institutes of Health, "Stress in poverty may impair learning ability in young children," National Institutes of Health: Turning Discovery into Health, 2013, accessed January 22, 2015, http://www.nih.gov/news/health/aug2012/nichd-28.htm.

114 *tools given them by their caretakers:* "Vygotskian approach: Lev Vygotsky," Tools of the Mind, 2015, accessed January 21, 2015, http://www.toolsofthemind.org/philosophy/vygotskian-approach/.

114 *problems related to self-regulation:* For the effects of language difficulties on the development of executive function, see Lucy A. Henry, David J. Messer, and Gilly Nash, "Executive functioning in children with specific language impairment," *Journal of Child Psychology and Psychiatry* 53.1 (2012): 37–45. For the impact of deafness on the development of executive function, see B. Figueras, L. Edwards, and D. Langdon, "Executive function and language in deaf children," *Journal of Deaf Studies and Deaf Education* 13.3 (2008): 362–377.

115 *both a child's language and his or her social skills:* Susan Hendler Lederer, "Efficacy of parent-child language group intervention for late-talking toddlers," *Infant-Toddler Intervention: The Transdisciplinary Journal* 11 (2001): 223–235.

115 *predicted improved social skills into early adolescence:* Michael D. Niles, Arthur J. Reynolds, and Dominique Roe-Sepowitz, "Early childhood intervention and early adolescent social and emotional competence: Second-generation evaluation evidence from the Chicago Longitudinal Study," *Educational Research* 50.1 (2008): 55–73.

115 *children from high-risk families:* Ibid.

115 *greater social skills and fewer behavioral problems:* Adam Winsler, J.R. De León, B.A. Wallace, M.P. Carlton, and A. Willson-Quayle, "Private speech in preschool children: Developmental stability and change, across-task consistency, and relations with classroom behaviour," *Journal of Child Language* 30.03 (2003): 583–608.

115 *Teachers rated these children higher for self-regulation:* Natalie Yvonne Broderick, "An investigation of the relationship between private speech and emotion regulation in preschool-age children," *Dissertation Abstracts International, Section B: The Sciences and Engineering* 61.11 (2001): 6125.

115 *negative outcomes for self-control and social skills:* Laura E. Berk and Ruth A. Garvin, "Development of private speech among low-income Appalachian children," *Developmental Psychology* 20.2 (1984): 271–286.

116 *mathematics that increased into the first grade:* Clancy Blair and C. Cybele Raver, "Closing the achievement gap through modification of neurocognitive and neuroendocrine function: Results from a cluster randomized controlled trial of an innovative approach to the education of children in kindergarten," *PLOS ONE* 9.11 (2014): e112393, doi:10.1371/journal.pone.0112393.

116 *"those in high-income school districts":* Clancy Blair, interview with the author, January 5, 2015.

116 *ability to regulate behavior and emotional responses:* Brian E. Vaughn, Claire B. Kopp, and Joanne B. Krakow, "The emergence and consolidation of self-control from eighteen to thirty months of age: Normative trends and individual differences," *Child Development* (1984): 990–1004.

117 *aspect of executive function and self-regulation:* Christopher M. Conway, David B. Pisoni, and William G. Kronenberger, "The importance of sound for cognitive sequencing abilities the auditory scaffolding hypothesis," *Current Directions in Psychological Science* 18.5 (2009): 275–279.

117 *an even more fundamental, more profound, level:* W.G. Kronenberger, J. Beer, I. Castellanos, D.B. Pisoni, and R.T. Miyamoto, "Neurocognitive risk in children with cochlear implants," *JAMA Otolaryngology—Head and Neck Surgery* (2014), doi:10.1001/jamaoto.2014.757; William G. Kronenberger, D.B. Pisoni, S.C. Henning, and B.G. Colson, "Executive functioning skills in long-term users of cochlear implants: A case control study," *Journal of Pediatric Psychology* 38.8 (2013): 902–914.

118 *notably stronger executive function and self-regulation:* Célia Matte-Gagné and Annie Bernier, "Prospective relations between maternal autonomy support and child executive functioning: Investigating the mediating role of child language ability," *Journal of Experimental Child Psychology* 110.4 (2011): 611–625.

118 *non-emotional reasons for discipline:* This is a rich and varied line of investigation in the developmental literature. For various discussions of the relationship between adult validation of child attempts at self-control and the development of self-regulation, see Grazyna Kochanska and Nazan Aksan, "Children's conscience and self-regulation," *Journal of Personality* 74.6 (2006): 1587–1618; Peggy Estrada, William F. Arsenio, Robert D. Hess, and Susan D. Holloway, "Affective quality of the mother-child relationship: Longitudinal consequences for children's school-relevant cognitive functioning," *Developmental Psychology* 23.2 (1987): 210–215; Robert C. Pianta, Sheri L. Nimetz, and Elizabeth Bennett, "Mother-child relationships, teacher-child relationships, and school outcomes in preschool and kindergarten," *Early Childhood Research Quarterly* 12.3 (1997): 263–280; Robert C. Pianta, Michael S. Steinberg, and Kristin B. Rollins, "The first two years of school: Teacher-child relationships and deflections in children's classroom adjustment," *Development and Psychopathology* 7.02 (1995): 295–312.

118 *these become the basis of their own behavior:* For further reading on how parent language shapes child use of private speech for self-regulation, see Rafael M. Diaz, A. Winsler, D.J. Atencio, and K. Harbers, "Mediation of self-regulation through the use of private speech," *International Journal of Cognitive Education and Mediated Learning* 2.2 (1992): 155–167; Adam Winsler, "Parent-child interaction and private speech in boys with ADHD," *Applied Developmental Science* 2.1 (1998): 17–39; Adam Winsler, Rafael M. Diaz, and Ignacio Montero, "The role of private speech in the transition from collaborative to independent task performance in young children," *Early Childhood Research Quarterly* 12.1 (1997): 59–79.

118 *negative impact of parents who are more controlling:* For further reading, see Annemiek Karreman, C.V. Tuijl, and A.G. Marcel, "Parenting, co-parenting, and effortful control in preschoolers," *Journal of Family Psychology* 22.1 (2008): 30–40; Grazyna Kochanska and Amy Knaack, "Effortful control as a personality characteristic of young children: Antecedents, correlates, and consequences," *Journal of Personality* 71.6 (2003): 1087–1112.

119 *a preponderance of directives impairs it:* Susan H. Landry, K.E. Smith,

P.R. Swank, and C.L. Miller-Loncar, "Early maternal and child influences on children's later independent cognitive and social functioning," *Child Development* 71.2 (2000): 358–375.

119 *emerging executive function and self-regulation skills:* See Alfred L. Baldwin, *Behavior and Development in Childhood* (Fort Worth, TX: Dryden Press, 1955); Claire B. Kopp, "Antecedents of self-regulation: A developmental perspective," *Developmental Psychology* 18.2 (1982): 199–214.

119 *some research finds that these children thrive:* For further reading, see Jay Belsky and Michael Pluess, "Beyond diathesis stress: Differential susceptibility to environmental influences," *Psychological Bulletin* 135.6 (2009): 885–908; W. Thomas Boyce and Bruce J. Ellis, "Biological sensitivity to context: I. An evolutionary-developmental theory of the origins and functions of stress reactivity," *Development and Psychopathology* 17.02 (2005): 271–301.

121 *negatively impacted intellectual development and IQ:* "Cognitive advantages of bilingualism," *Wikipedia*, Wikipedia Foundation, June 9, 2014, accessed January 22, 2015, http://en.wikipedia.org/wiki/Cognitive _advantages_of_bilingualism.

121 *revealed in 1962 by Professors Elizabeth Peal and Wallace Lambert:* Elizabeth Peal and Wallace E. Lambert, "The relation of bilingualism to intelligence," *Psychological Monographs: General and Applied* 76.27 (1962): 1–23.

121 *Ellen Bialystok:* Quoted in Alexandra Ossola, "Are bilinguals really smarter?: Despite what you may have read, it's not so cut and dry," *Science Line,* July 29, 2014, accessed January 22, 2015, http://science line.org/2014/07/are-bilinguals-really-smarter/.

123 *vocabulary, syntax, nuance, or overall quality:* For examples of Professor Hoff's extensive research in this area, see Erika Hoff, R. Rumiche, A. Burridge, K.M. Ribot, and S.N. Welsh, "Expressive vocabulary development in children from bilingual and monolingual homes: A longitudinal study from two to four years," *Early Childhood Research Quarterly* 29.4 (2014): 433–444; Silvia Place and Erika Hoff, "Properties of dual language exposure that influence 2-year-olds' bilingual proficiency," *Child Development* 82.6 (2011): 1834–1849.

123 *impacts overall cognitive development at twenty-four months:* Negative impacts of non-native language on cognitive development were reflected in lower Bayley Scales intelligence scores at twenty-four months of age. Adam Winsler, Margaret R. Burchinal, Hsiao-Chuan Tien, Ellen Peisner-Feinberg, Linda Espinosa, Dina C. Castro, Doré R. LaForett, Yoon Kyong Kim, and Jessica De Feyter, "Early development among

dual language learners: The roles of language use at home, maternal immigration, country of origin, and socio-demographic variables," *Early Childhood Research Quarterly* (2014): 750–764.

124 **Give and Take: Why Helping Others Drives Our Success:** Adam Grant, *Give and Take: A Revolutionary Approach to Success* (New York: Viking, 2013).

124 *"Raising a Moral Child":* Adam Grant, "Raising a moral child," *The New York Times,* April 11, 2014, accessed January 22, 2015, http://www.ny times.com/2014/04/12/opinion/sunday/raising-a-moral-child.html?_r=0.

125 *more likely to be generous:* Ibid.

125 *children who had heard nothing:* Ibid.

Chapter 5: The Three Ts

137 *less likely to learn the words being used:* G. Hollich, K. Hirsh-Pasek, and R.M. Golinkoff, "Breaking the language barrier: An emergentist coalition model for the origins of word learning," *Monographs of the Society for Research in Child Development* 65.3, serial no. 262 (2000).

138 *the Middle East, and Australia:* Anne Fernald and Patricia Kuhl, "Acoustic determinants of infant preference for motherese speech," *Infant Behavior and Development* 10 (1987): 279–293; A. Fernald, T. Taeschner, J. Dunn, M. Papousek, B. de Boysson-Bardies, and I. Fukui, "A cross-language study of prosodic modifications in mothers' and fathers' speech to preverbal infants," *Journal of Child Language* 16.3 (1989): 477–501; A. Kelkar, "Marathi baby talk," *Word* 20 (1965): 40–54; P.B. Meegaskumbura, "Tondol: Sinhala baby talk," *Word* 31.3 (1980): 287–309; Nobuo Masataka, "Motherese in a signed language," *Infant Behavior and Development* 15.4 (1992): 453–460.

138 *during a two-week home-visiting program:* P.W. Jusczyk and E.A. Hohne, "Infants' memory for spoken words," *Science* 277.5334 (1997): 1984–1986.

139 *Science tells us clearly that:* For discussions of this process, see Engle and Ricciuti, "Psychosocial aspects of care and nutrition"; C.M. Heinicke, N.R. Fineman, G. Ruth, S.L. Recchia, D. Guthrie, and C. Rodning, "Relationship-based intervention with at-risk mothers: outcomes in the first year of life," *Infant Mental Health Journal* 20 (1999): 249–274; N. Eshel, B. Daelmans, M. Cabral de Mello, and J. Martines, "Responsive parenting: interventions and outcomes," *Bulletin of the World Health Organization* 84 (2006): 992–999.

139 *essential to behavioral and brain development:* C.S. Tamis-LeMonda and M.H. Bornstein, "Habituation and maternal encouragement of

attention in infancy as predictors of toddler language, play, and representational competence," *Child Development* 60 (1989): 738–751.

139 *1. Observation 2. Interpretation 3. Action:* For discussions of effective parent responsiveness, see J.P. Shonkoff and D.A. Phillips, eds., *From Neurons to Neighborhoods: The Science of Early Child Development* (Washington, DC: National Academy Press, 2000); L. Richter, *The Importance of Caregiver-Child Interactions for the Survival and Health Development of Young Children: A Review* (Geneva: World Health Organization, 2004); P.L. Engle and H.N. Riccituti, "Psychological aspects of care and nutrition," *Food and Nutrition Bulletin* 16 (1995): 356–377.

141 *autoimmune disorders:* G. Miller and E. Chen "Unfavorable socioeconomic conditions in early life presage expression of proinflammatory phenotype in adolescence," *Psychosomatic Medicine* 69.5 (2007): 402–409.

147 *significant relationship to a child's brain development:* R. Paul, *Language Disorders from Infancy Through Adolescence,* 2nd ed. (St. Louis, MO: Mosby, 2001).

151 *Terry Paul:* LENA Research Foundation, "Our story," accessed February 26, 2015, http://www.lenafoundation.org/about-us/founders -story/. For further reading on pretend play, see G.S. Ashiabi, "Play in the preschool classroom: Its socioemotional significance and the teacher's role in play," *Early Childhood Education Journal* 35 (2007): 199–207; L.E. Berk, T.D. Mann, and A.T. Ogan, "Make-believe play: Wellspring for development of self-regulation," in *Play = Learning: How Play Motivates and Enhances Children's Cognitive and Social-Emotional Growth,* ed. D. Singer, R.M. Golinkoff, and Hirsh-Pasek (New York: Oxford University Press, 2006); J.F. Jent, L.N. Niec, and S.E. Baker, "Play and interpersonal processes," in *Play in Clinical Practice: Evidence-Based Approaches,* ed. S.W. Russ and L.N. Niec (New York: Guilford Press, 2011); S.W. Russ, *Play in Child Development and Psychotherapy* (Mahwah, NJ: Lawrence Erlbaum Associates, 2004); A.L. Seja and S.W. Russ, "Children's fantasy play and emotional understanding," *Journal of Clinical Child Psychology* 28 (1999): 269–277.

152 *when he read Hart and Risley's* **Meaningful Differences:** See chap. 2 notes for complete citation information for this work.

154 *being a more successful reader:* L. Baker, R. Serpell, and S. Sonnenschein, "Opportunities for literacy learning in the homes of urban preschoolers," in *Family Literacy Connections in Schools and Communities,* ed. L. Morrow (Newark, NJ: IRA, 1995), 236–252; C. Snow and P. Tabors, "Intergenerational transfer of literacy," in *Fam-*

ily Literacy: Directions in Research and Implications for Practice, ed. L.A. Benjamin and J. Lord (Washington, DC: Office of Education Research and Improvement, U.S. Department of Education, 1996).

159   *read to without gesturing:* S.B. Piasta, L.M. Justice, A.S. McGinty, and J.N. Kaderavek, "Increasing young children's contact with print during shared reading: Longitudinal effects on literacy achievement," *Child Development* 83.3 (2012): 810–820.

159   *future vocabulary in their children:* C. Peterson, B. Jesso, and A. McCabe, "Encouraging narratives in preschoolers: An intervention study," *Journal of Child Language* 26.1 (1999): 49–67.

165   *technology, engineering, and math fields:* S. Franceschini, S. Gori, M. Ruffino, K. Pedrolli, and A. Facoetti, "A causal link between visual spatial attention and reading acquisition," *Current Biology* 22.9 (2012): 814–819; Jonathan Wai, David Lubinski, and Camilla P. Benbow, "Spatial ability for STEM domains: Aligning over 50 years of cumulative psychological knowledge solidifies its importance," *Journal of Educational Psychology* 101.4 (2009): 817–835.

171   *jump start on learning:* Deborah Stipek, "Q&A with Deborah Stipek: building early math skills," Stanford University Graduate School of Education, https://ed.stanford.edu/in-the-media/qa-deborah-stipek -building-early-math-skills, accessed March 4, 2015.

191   *human-to-human interaction:* R. Barr and H. Hayne, "Developmental changes in imitation from television during infancy," *Child Development* 70.5 (1999): 1067–1081.

Chapter 6: The Social Consequences

194   *inequality in the United States over the past four decades:* Cf. Sean F. Reardon, "No rich child left behind," *The New York Times*, April 27, 2013, accessed February 17, 2015, http://opinionator.blogs.nytimes .com/2013/04/27/no-rich-child-left-behind/.

194   *To me, this is both admirable and important:* S.F. Reardon, "The widening academic achievement gap between the rich and the poor: New evidence and possible explanations," in *Whither Opportunity? Rising Inequality, Schools, and Children's Life Chances*, ed. Greg J. Duncan and Richard J. Murnane (New York: Russell Sage Foundation, 2011).

197   *the key tool to make the necessary changes:* Steve Dow, personal interview with the author, February 9, 2015.

198   *While parenting interventions:* Greg J. Duncan and Richard J. Murnane, "Introduction: The American dream, then and now," in Duncan and Murnane, *Whither Opportunity?*, pp. 3–26.

198   **Unequal Childhoods:** Annette Lareau, *Unequal Childhoods: Class, Race and Family Life* (Berkeley: University of California Press, 2003), 343.

199   *"neither equal nor freely chosen":* Ibid.

199   *"realistic picture of the day-to-day rhythms of family":* Annette Lareau, "Question and Answers: Annette Lareau, *Unequal Childhoods: Class, Race, and Family Life;* University of California Press," 2003, p. 1, accessed February 13, 2015, https://sociology.sas.upenn.edu/sites /sociology.sas.upenn.edu/files/Lareau_Question&Answers.pdf.

199   *her team said they wanted to be like "the family dog":* See Lareau, *Unequal Childhoods,* p. 9, citing Ariel Hochschild in Ariel Hochschild and Anne Machung, *The Second Shift: Working Parents and the Revolution at Home* (New York: Avon, 1989).

199   *"allow us to hang out with them":* Ibid., p. 9.

199   *beauty parlors, barbershops, and even staying overnight:* Lareau, "Question and Answers," p. 1.

200   *"children to be happy and to grow and thrive":* Ibid.

200   *a kind of frenetic energy:* Lareau, *Unequal Childhoods,* p. 5.

200   *"more familiarity with abstract concepts":* Ibid.

200   *"verbal jousting" and "word play":* Quoted material from a summary of Lareau's *Unequal Childhoods* in Linda Quirke, "Concerted Cultivation / Natural Growth," in *Sociology of Education: An A-to-Z Guide,* ed. James Ainsworth (Los Angeles: Sage Publications, 2013), pp. 143–145.

200   *"except in matters of health or safety":* Ibid.

200   *"the accomplishment of 'natural growth'":* Lareau, *Unequal Childhoods,* quote on p. 3.

201   *being handed a washcloth:* Ibid., p. 147.

202   *getting to the goals that was the difference:* Ibid., p. 386.

202   *social scientists in our democratic future:* A. Lareau, "Cultural knowledge and social inequality," *American Sociological Review* 80.1 (2015): 1–27.

204   *seven-and-a-half-year-old children:* Elizabeth A. Moorman and Eva M. Pomerantz, "Ability mindsets influence the quality of mothers' involvement in children's learning: An experimental investigation," *Developmental Psychology* 46.5 (2010): 1354–1462.

206   *being hit when you are down:* Ibid.

208   *"understanding of the world at that time":* Portia Kennel, personal communication, January 16, 2015.

210   *Wes Moore:* "The other Wes Moore? Expectations matter," *Idea Festival,* accessed February 13, 2015, http://www.ideafestival.com/index .php?ption=com_contentandview=articleandid=10692:who -is-the-other-wes-moorandcatid=39:if-blog.

213 *"better invest in their children":* A. Kalil, "Inequality begins at home: The role of parenting in the diverging destinies of rich and poor children," in *Diverging Destinies: Families in an Era of Increasing Inequality,* eds. P. Amato, A. Booth, S. McHale, and J. Van Hook (New York: Springer, 2014), pp. 63–82.

217 *costing billions and billions, have little or no effect:* Ron Haskins, "Social programs that work," *The New York Times,* December 31, 2014, accessed March 3, 2015, http://www.nytimes.com/2015/01/01 /opinion/social-programs-that-work.html?_r=0.

218 *"this essential R&D dimension":* Center for High Impact Philanthropy, University of Pennsylvania, "Investing in Early Childhood Innovation: Q&A with Dr. Jack P. Shonkoff," March 30, 2015, http://www.impact .upenn.edu/2015/03/investing-in-early-childhood-innovation-qa-with -dr-jack-p-shonkoff/, accessed March 3, 2015.

220 *calls the "twin streams":* Ellen Galinsky, personal communication, January 29, 2015.

221 *clues for how it might have been a striking success:* P. Lindsay Chase-Lansdale and Jeanne Brooks-Gunn, "Two-generation programs in the twenty-first century," *Future of Children* 24.1 (2014): 13–39, accessed February 18, 2015, http://futureofchildren.org/futureofchildren/ publications/docs/24_01_FullJournal.pdf.

221 *supportive parent and breadwinner:* Carolyn J. Heinrich, "Parents' employment and children's wellbeing," *Future of Children* 24.1 (2014): 121–146, accessed February 18, 2015, http://futureofchildren.org/future ofchildren/publications/docs/24_01_FullJournal.pdf.

221 *the Community Action Project (CAP) of Tulsa:* To learn more about CAP Tulsa, visit the organization's website, accessed March 3, 2015, https://captulsa.org/.

Chapter 7: Spreading the Words

224 *What makes us want to be part of encouraging it?:* Atul Gawande, "Slow ideas," *The New Yorker,* July 29, 2013, accessed March 4, 2015, http://www.newyorker.com/magazine/2013/07/29/slow-ideas.

225 *The achievement gap is already present at nine months of age:* Tamara Halle, Nicole Forry, Elizabeth Hair, Kate Perper, Laura Wandner, Julia Wessel, and Jessica Vick, "Disparities in early learning and development: Lessons from the early childhood longitudinal study— birth cohort (ECLS-B)," *Child Trends* (2009), accessed February 23, 2015, http://www.childtrends.org/wp-content/uploads/2013/05/2009 -52DisparitiesELExecSumm.pdf.

227 *accounting for 28 percent of the world's total:* US Energy Information Administration, "United States leads world in coal reserves," *Today in Energy* (2011), accessed February 18, 2015, http://www.eia.gov/today inenergy/detail.cfm?id=2930.

227 *It is one of the world's largest national economies:* The United States had the largest economy in the world until the end of 2014, when China's economy surpassed the United States', $17.6 trillion to $17.4 trillion, according to the International Monetary Fund (IMF). See Ben Carter, "Is China's economy really the largest in the world?" *BBC News Magazine,* 2014, accessed February 18, 2015, http://www.bbc .com/news/magazine-30483762.

228 *sixteen million of whom lived below the poverty line:* Yang Jiang, Mercedes Ekono, and Curtis Skinner, "Basic facts about low-income children: Children under 18 Years, 2013," National Center for Children in Poverty, Mailman School of Public Health, Columbia University, January 1, 2015, accessed March 4, 2015, http://www.nccp.org /publications/pdf/text_1100.pdf.

229 *healthy behavior, and adult productivity:* James J. Heckman, "The economics of inequality: The value of early childhood education," *American Educator* (2011), accessed February 18, 2015, https://www .aft.org//sites/default/files/periodicals/Heckman.pdf.

230 *"Bridging the Thirty-Million-Word Gap":* See the Bridge the Word Gap website at https://bridgethewordgap.wordpress.com for an overview of the meeting results and a list of participating agencies and organizations.

230 *"Public Policies, Made to Fit People":* Richard Thaler, "Public Policies, Made to Fit People," *The New York Times,* August 24, 2013, accessed March 4, 2015, http://www.nytimes.com/2013/08/25/business /public-policies-made-to-fit-people.html?_r=0>.

230 *awarded the grand prize by the Bloomberg Mayors Challenge:* "2012–2013 Mayors Challenge," Bloomberg Philanthropies: Mayors Challenge, accessed March 4, 2015, http://mayorschallenge.bloomberg.org /index.cfm?objectid=7E9F3B30-1A4F-11E3-8975000C29C7CA2F.

231 *"Bridging the Early Language Gap: A Plan for Scaling Up":* Dana Suskind, Patricia Kuhl, Kristin R. Leffel, Susan Landry, Flávio Cunha, and Kathryn M. Nevkerman, "Bridging the early word gap: A plan for scaling up" (white paper prepared for the White House meeting on Bridging the Thirty-Million-Word Gap), September 2013.

235 *Dr. Perri Klass:* Personal communication, February 19, 2015.

235 *fee-for-service world of medicine:* Carol Peckham, "Number of patient visits per week," slide 17, *Medscape Pediatrician Compensation Re-*

port 2014 (2014), accessed February 18, 2015, http://www.medscape
.com/features/slideshow/compensation/2014/pediatrics#17.

240 *Professors Noshir Contractor and Leslie DeChurch:* Noshir S. Contractor and Leslie A. DeChurch, "Integrating social networks and human social motives to achieve social influence at scale," *Proceedings of
the National Academy of Sciences of the United States of America*
111.4 (2014): 13650–13657.

241 *"scientific discoveries to public good":* Ibid., p. 13650.

241 *"beliefs and norms in the minds of the many":* Ibid.

241 *attitude change and new norms within the community:* Ibid., p. 13655.

Appendix: Early Childhood Organizations and Resources

251 *Too Small To Fail:* Learn more about Too Small to Fail at http://too
small.org/.

252 *Talk With Me Baby:* Learn more about Talk With Me Baby at http://
www.talkwithmebaby.org/.

252 *Reach Out And Read:* Learn more about Reach Out and Read at http://
www.reachoutandread.org/.

253 *who were not in Reach Out and Read in their preschool years:* "Research
Findings," Reach Out and Read, accessed February 23, 2015, http://www
.reachoutandread.org/why-we-work/research-findings/. This link provides a complete listing of published research conducted by Reach Out
and Read.

253 *Educare:* To learn more about Educare, visit http://www.educare
schools.org/locations/chicago.php.

253 *Mind In The Making:* Learn more about Mind in the Making at http://
www.mindinthemaking.org/.

254 *Vroom:* Learn more about Vroom at http://www.joinvroom.org/.

255 *Providence Talks:* To learn more about Providence Talks, visit http://
www.providencetalks.org/.

ACKNOWLEDGMENTS

Thirty Million Words is a reflection of the incredible, nonstop, never-accept-anything-but-the-best team that has developed a multilayered, finely honed research project from a tiny, embryonic idea that we could somehow help children at risk succeed. This book is reflective of their work and their dedication. Kristin Leffel and Beth Suskind, one of the coauthors of this book, have been with TMW almost from its inception. Their humanity, creativity, brilliance, and stalwart support have been invaluable. As TMW grows, our family grows, each member with a different expertise, yet all with the same creative, intellectual energy that drives TMW's excellence: Eileen Graf, Ashley Telman, Iara Fuenmayor, Tara Robinson, Alison Hundertmark, Rachel Umans, Sarah Van Deusen Phillips, Livia Garofalo, Alyssa Anneken, and Macarena Galvez. Our extended TMW family includes Marc Hernandez, Karen Skalitzky, Sally Tannenbaum, Michelle Havlik, Lydia Polonsky, Mary Ellen Nevins, Shannon Sapolich, Debbie Hawes, Lyra Repplinger, Andrea Rohlfing, Hannah Bloom, and Karen Pekow. And the incredible undergraduate and graduate research assistants who keep the bustling lab afloat. I am indebted to you all. You make me look good!

The funders of Thirty Million Words are true partners and dear friends. The Hemera Foundation believed in TMW and supported the vision from the beginning. Thank you, Caroline Pfohl, for bringing extraordinary mindfulness to the forefront and, of course, for "Tuning In." Thank you, Rob Kaufold, for always being a great support and a true partner. Thank you, Rick White, for being a solid foundation, always making sure I never take myself too seriously. Thank you, Rebecca White, for your extraordinary optimism and Jay Hughes, thank you, thank you for reading Hart and Risley! Simply saying "thank you," in fact, is not enough to express my gratitude. You allowed the Thirty Million Words initiative to get its roots, and we wouldn't be where we are today without you.

Thank you, too, to the PNC Grow Up Great Foundation, the W.K. Kellogg Foundation, the Robert R. McCormick Foundation, and the Hyman Milgrom Supporting Organization. You have all ensured that this innovative science continues.

The University of Chicago, the University of Chicago Medicine, and the Institute for Translational Medicine are my home. I thank every person and department for so solidly and enthusiastically supporting a surgeon with such a crazy idea. Thank you, Jeff Matthews, for the seed money that started it all.

It's been an extraordinary experience to be an absolute neophyte and still warmly supported by so many experts in the field who could have rolled their eyes at a surgeon daring to venture out of the operating room. Instead, they have been wonderful guides, sharing their expertise and their brilliant analysis generously. Thank you, also, Susan Levine and Susan Goldin-Meadow, for educating me from the beginning.

Thank you, too, to everyone who's taken time from busy schedules to give me invaluable, constructive feedback: Cornelia Grumman, Liz Gunderson, Clancy Blair, Kavita Kapadia, Debbie Leslie, Shayne Evans, Steve Dow, Ann Hanson, Tony Raden, Portia Kennel, Diana Rauner, Megan Roberts, Ariel Kalil, Ellen Galinsky, Kathy Hirsh-Pasek, Jack Shonkoff, and many others.

To my agent, Katinka Matson, who believed in us and found exactly the right match for this idea. To Stephen Morrow, for seeing the potential and helping make that potential become a reality.

To all the amazing TMW parents from whom I have learned so much. May this book honor in some small way your strength, love, and dedication. We have much more work to do, and I'm so happy to be doing it alongside each of you.

Most important, to my wonderful, supportive family, which has been with me every step of the way, even when the steps were very hard to climb: Michael, Beth, Sydnie, Yonah, David, Rebecca, Lily, Carter, Noa, Emmett, Elias, and Sadie. And of course, Lola and Nghia. Thank you to my wonderful and loving parents, Bob and Leslie. Especially for all the process-based praise! Mom, we've given new meaning to book sharing and bonding. Thank you so much for Tuning In . . . and Taking Turns!

Above all, to my incredible children, Genevieve, Asher, and Amelie, who inspire me every day and without whose support I could have never finished this book. Now that it's done, I promise to begin Talking More . . . and maybe even Taking Turns! I love you.

INDEX

Note: The abbreviation "TMW" denotes "Thirty Million Words."